Sky and Earth

Giuseppe Mussardo · Gaspare Polizzi

Sky and Earth

Travelling with Dante Alighieri and Marco Polo

Giuseppe Mussardo
International School for Advanced Studies
Trieste, Italy

Gaspare Polizzi
University of Pisa
Pisa, Italy

Drawings in the book by Tamari Mirotadze
© 2021 Edizioni Dedalo
divisione della Dedalo Litostampa srl

ISBN 978-3-031-18610-3   ISBN 978-3-031-18611-0 (eBook)
https://doi.org/10.1007/978-3-031-18611-0

Translation from the Italian language edition: "Tra cielo e terra. In viaggio con Dante Alighieri e Marco Polo" by Giuseppe Mussardo and Gaspare Polizzi, © Dedalo 2021. Published by edizioni Dedalo. All Rights Reserved.
The translation was done with the help of artificial intelligence (machine translation by the service DeepL.com). A subsequent human revision was done primarily in terms of content.
© The Editor(s) (if applicable) and The Author(s), under exclusive license to Springer Nature Switzerland AG 2023
This work is subject to copyright. All rights are solely and exclusively licensed by the Publisher, whether the whole or part of the material is concerned, specifically the rights of translation, reprinting, reuse of illustrations, recitation, broadcasting, reproduction on microfilms or in any other physical way, and transmission or information storage and retrieval, electronic adaptation, computer software, or by similar or dissimilar methodology now known or hereafter developed.
The use of general descriptive names, registered names, trademarks, service marks, etc. in this publication does not imply, even in the absence of a specific statement, that such names are exempt from the relevant protective laws and regulations and therefore free for general use.
The publisher, the authors, and the editors are safe to assume that the advice and information in this book are believed to be true and accurate at the date of publication. Neither the publisher nor the authors or the editors give a warranty, expressed or implied, with respect to the material contained herein or for any errors or omissions that may have been made. The publisher remains neutral with regard to jurisdictional claims in published maps and institutional affiliations.

This Springer imprint is published by the registered company Springer Nature Switzerland AG
The registered company address is: Gewerbestrasse 11, 6330 Cham, Switzerland

Prologue

According to legend, Dante Alighieri and Marco Polo met in Verona in 1313, during a court banquet. It had been more than ten years since the Poet had been sent into exile from his equally beloved and hated Florence: he had wandered through various regions of Italy, but, for about a year, he had found a generous welcome with Cangrande della Scala, lord of Verona, who was both a shrewd leader and a skilful politician. Many years before, Marco Polo had also been generously welcomed by a "Gran Can," or, rather, by Kublai Khan, lord of all the Mongols, whose immense empire stretched from the shores of the Pacific Ocean to those of the Black Sea, from the swampy steppes of cold Siberia to the white Himalayas and the forests of Burma.

What the Poet and the Merchant said to each other, these two great storytellers, the first in rhyme, the second in prose, in an Italian language that was beginning to take shape thanks to the two of them, remains unknown to this day. If politics had been Dante's asphyxiating passion, a passion that had consumed him to the bone and for which he was paying a very high price, geography and the customs and traditions of the various peoples of the Far East, or, rather, a curiosity about all that was new and exotic, had been the driving force that had pushed Marco Polo to cross deserts, climb the peaks of the Pamir, and travel across the immense plains of Asia.

Both had travelled to unknown places, in a great race towards the unknown: they had followed the rhythm of the waves, the storm of the winds; one had been intoxicated by the vastness of immense landscapes, the other by dark circles of hell. Both had experienced the frost of the boundless frozen swamps and the blinding light of the Sun, that star that guides and moves all the other stars. Each had always said, "Let's go on, I want to see what's beyond," one with a heart full of rancour, the other with the lightness of a soap bubble.

They had been interested in chess and angels, sinners and brigands, saints and murderers, astonishing miracles and great iniquities, empires and religions, majestic rivers and equally immense seas, as well as astrolabes and compasses, earth and sky, geography and cosmogony.

Their love and thirst for knowledge made them take off on a "mad flight," one towards the unexplored territories of Asia, the other towards the remote areas of history and the mind. Two extraordinary travellers who are worth following, whether in their footsteps, in the noble stories they have to tell, or in the formidable memories of which their treasure chests are full.

Comparing and interweaving the stories of these two journeys, towards the immensity of the divine and along the vague and undefined spaces of unknown lands, leads us to look, as if through a fascinating kaleidoscope, at a world far away in time and space that, after seven centuries, still causes us to dream and satisfies our thirst for knowledge. If, in Dante's *Commedia*, every tale is allegorical, in Marco Polo's *Il Milione*, every tale, when it is not an explicit novel, has instead the photographic cut of reportage.

James Joyce said: "I love my Dante as much as I love the Bible, he is spiritual food." Honoré de Balzac tried, in his Human Comedy, to render the same wealth of characters and situations as Dante's poem. The Russian poet Osip Mandel'štam never left his apartment in Moscow without a copy of the Comedy in case he was arrested, which he was, more than once. About Marco Polo, the famous sinologist Jonathan Spence wrote: "It's a strange disease. It can strike at any time. The symptomatology is clear enough: there is an overwhelming fascination with everything Marco has said and described. The cure is unknown."

Dante Alighieri and Marco Polo, two sides of a single medal, one celebrating fourteenth-century Italy, a time and place that will never cease to amaze us with its bloody richness.

Trieste, Italy Giuseppe Mussardo
Pisa, Italy Gaspare Polizzi

Contents

The Merchant and the Poet	1
Ordinary Geniuses and Sorcerers	2
The Giant Crystal	5
The Enigma of the Lost Manuscript	12
The Disease of Writing	14
In the Darkness of Hell	18
The Hill of Purgatory	27
The Light of Paradise	34
The Torments of Botticelli	37
A Non-existent Book	40
Diplomatic Missions	47
The Art of Travelling	49
The Never-Ending Querelle	57
Venice and Florence	61
At the Dawn of the Thirteenth Century	62
The World Seen from Rialto	64
An Extraordinary Republic	68
The Horses of San Marco	72
The Art of Wool	75
At the Roots of Hatred	80
No Holds Barred	81
International Intrigue	86
How Salty Other People's Bread Tastes	96

Astrolabe and Compass — 103
It All Began in Jerusalem — 104
The Shadow of an Ancient Master — 106
Behind the Scenes — 109
In the Heart of Tartary — 110
The Noble Castle — 113
Pax Mongolica — 118
Francesca and the Others, or Love and Sin — 121
Women of the Orient — 125
Monsters, Animals and Giants — 127
The Ineffable Priest Gianni — 134
Travelling in a Stubborn and Contrary Direction — 137
The Sect of Assassins — 140
Southern Cross — 144
In the Footsteps of Alexander the Great — 148
The Fire Purgatory — 153
The Miracle of the Mountain — 156
Dreams, Prayers and Numbers — 158
Round Trip from Ormuz — 164
Flowers and Trees, Light and Fire — 166
The Highest Place in the World — 169
The Elements, the Ether and Transhumanar — 173
The Silk Road — 175
Light, Dance and Music — 179
At the Court of the Great Khan — 185
The Eagle Question — 191
The Discovery of Asia — 193
Medieval Genius — 196
La Candida Rosa — 198
The Long Goodbye — 202

Recommended Reading — 207

The Merchant and the Poet
In Which We Learn of an "Ordinary Genius", a "Sorcerer", a Gigantic Crystal, a Genoese Prison and Two Non-Existent Books

Ordinary Geniuses and Sorcerers

Mark Kac, a talented mathematician of Polish origin, as well as a great scholar of random laws and the theory of probability that governs them, in his brilliant autobiography *The Enigmas of Chance*, ventured to argue that, in science, as in all other fields, there are two types of geniuses, what he called "ordinary geniuses" and "sorcerers".

In his words, "an ordinary genius is like one of us, and we could be like him if only we were many times better. There is no mystery as to how his mind works: seeing the result, we feel that surely we too would have been able to reach it." With sorcerers, however, the matter is quite different. Sorcerers simply live in another world and "their mental processes are completely incomprehensible to us. We may understand the final result, but how they got there will always remain a mystery." They are like Philip K. Dick's travellers, inhabitants of unknown worlds thrown into our midst so that, thanks to them, fantasy can surpass reality, precede it and build upon it.

To corroborate his thesis of the two types of geniuses, Kac mentioned two names that have left a strong imprint on the history of twentieth century science: Hans Bethe, as the greatest example of ordinary genius, and Richard Feynman, as the undisputed champion of the sorcery of physics. These names may not mean much to most people, but, just to give an idea as to how prominent they are, it is worth remembering that Bethe, as well as being the first to understand the nuclear laws at the basis of the life of the stars and the Sun, directed the theoretical division of the Manhattan project that led to construction of the first atomic bomb, while Feynman, on the basis of "witch doctor" processes, was the first to understand quantum electrodynamics, or the way in which light interacts with electrons to give life to the infinite variety of the world that surrounds us.

In our case, if Marco Polo needs a separate discourse—which we shall get to shortly—before attributing to him the designation of "ordinary genius", there is, however, no doubt that Dante Alighieri was rather a "sorcerer": how else can we define his prodigious ability to forge, in his own image and likeness, a linguistic magma such as the fourteenth-century vernacular to produce a work of 14,233 endecasyllabic verses of complete meaning? And, just to give you the sort of figures that can sometimes help to solidify the titanic nature of a work in short order, his Comedy (as it is known, being a poem in three Cantiche, subsequently divided into thirty-three Cantos each, except for the first, the *Inferno*, which also has a preamble) has a total of 101,698 words, with an average of almost 33,900 words per Cantica and almost 1017 per Canto. One is stunned by this work of witchcraft that is the *Commedia*.

Obviously, it is not only a question of numbers: the sorcerer genius is the one who is not satisfied with discovering how the world works—they want to create one of their own at any cost.

The illness of the sorcerer-genius is the thirst for infinity that animates them, it is the cyclopean effort to remove every memory of our weakness, to announce to us that we will ascend instead towards the heavens, towards the empyrean of knowledge, into that circle of light where the keys to the world can be found. From birth, the sorcerer knows that they are predestined, chosen from above (or selected only by the laws of chance...) for the task of overcoming every limit, of launching themselves beyond the Herculean columns of knowledge, of throwing themselves into the waves of an endless sea, at the boundaries of space and time. As far as Dante is concerned, his gift of imposing himself, of fascinating us, of making us experience even the smells of the places through which he passes, has something superbly divine about it, a gift of which he seems perfectly aware: he feels like a scribe of God, a being directly in contact with the Prime Mover of the world, who speaks and writes as if under his direct dictation "that matter on which I am made scribe."

A very strong wind stirs the pages of the Comedy, animating them all and making them all shine between our fingers. Inside, we find ancient history, with its legends, love and hate, loyalty and betrayal, forgiveness and revenge, together with politics, astronomy, faith, reason and literature. To the human passions, in their infinite conjugations, is added the history of the world. Before our eyes, thousands of characters come to life, in a continuous metamorphosis that mixes real and imaginary protagonists: Farinata degli Uberti is as real as the demonic monster Geryon, poor Pier delle Vigne's dung, groaning and squeaking, competes in realism with the snarling voice and very long tail of Minos, the light of Beatrice's smile with the silence of the blessed, Ulysses' mad flight ends with a thud as much as Alichino's and Calcabrina's turns, the green lawn of the noble castle where mankind's greatest scholars stay is as tangible as the myriad of lights of the Candida Rosa. After Dante, Inferno, Purgatory and Paradise became part of us—the journey that changed his life changed ours forever.

If literature, as Jorge Luis Borges said, is nothing more than a guided dream, few other works leave the impression of a secret that escapes like the Comedy does, of a bazaar full of all of the stones of creation, from the vilest to the noblest, and of treasure chests that one never finishes opening. As well as being its author, Dante is also the undisputed protagonist and witness of that great dream that is the Comedy, in a narrative circularity taken up in modern times only by two great writers of the twentieth century, Marcel Proust, with his *Search for Lost Time*, another book that reveals itself by being written, and James Joyce, with *Ulysses*, the story of an equally extraordinary journey into the world of the universe, in the pubs, in the streets and guts of a magmatic Dublin, another brilliant crossing between real and imagined worlds.

The case of Marco Polo is different: although he tells us of witchcraft whose echoes come from the distant frontier places of medieval Asia, a world always suspended between reality and legend—salamanders that lived in fire, arrows that stopped in the air, deserts animated by talking sands—Marco Polo is anything but a sorcerer. His is the slow and compassionate voice of a traveller accustomed to both the silence of the mountains and the frenzy of the caravanserais, his is like the gaze of a great photographer who notices everything and judges nothing, whether the eye falls on astonishing natural landscapes or equally astonishing human customs. And it is still his voice that made the Silk Road known to the West, that ramified web of mule tracks, of very long itineraries, of impervious passages between the highest mountains in the world and the most desolate deserts on Earth where the possibility of doing business arose at every stop. Even the emperor of the Mongols learned to know the timbre of that voice when, at night, around the smoking braziers, the Venetian told of all the lands and peoples that the Khan had conquered, but had never had the chance to visit.

If we were to elevate geography, the "writing of the Earth" of the ancient Greeks, to the rank of science, well, Marco Polo would certainly be the undisputed "ordinary genius" of this science: scrolling through the pages of *Il Milione*, also known as *Le Divisament dou Monde*, everyone is led to believe that they are capable of crossing the fearsome Taklamakan desert or the snowy 7000-m peaks of the Afghan Pamir or of traversing the Indian Ocean with the same lightness and ease that transpires in Marco's tale.

On the other hand, if there is an absolute genius of understatement—that ever-so-British art of minimizing, of stating more by saying less—it is Marco Polo himself. Every writer writes in their mother tongue (with the possible exception of the singular case of Joseph Conrad), but their true language is very often another. Just to give an example, James Joyce could have been the Italian writer Carlo Emilio Gadda: if he had set his *Ulysses* in Rome, we would have had the voices and tones of a thousand Italian dialects, the Roman student, the Neapolitan postman, the Florentine notary, the journalist from Trieste, the Piedmontese commissioner, the Venetian librarian, and so on, and the book would perhaps have benefitted. Well, for Marco Polo, there is no doubt that his true language, the natural idiom of *Il Milione*, is precisely the British language: a short, essential touch, which disdains useless phrases, where the understatement of the text is equal to the author's lifestyle. The conciseness words, their conciseness and clarity, however, refer to stories with a much broader and dramatic scope than what the author seeks to portray: how can one not remain open-mouthed when one arrives at the passage in which Marco describes how he, his father Niccolò and his uncle Matteo left China by sea after staying in those lands for almost seventeen years? Here is the text:

> *I should add that in that expedition more than six hundred had boarded ships persons not counting the sailors: and all were dead except the eighteen survived. Of the three ambassadors only the one called Cogia was saved; and of the hundred women on board, only one remained alive.*

Six hundred dead, and not a word of explanation?! This, here, was Marco Polo… The woman who remained alive, by the way, was the reason for their last mission for Kublai Khan: the beautiful princess Cocacin, entrusted to the Polos so they could accompany her to Persia to meet up with Argon Khan, who had chosen her as his bride.

Il Milione, wrote Luigi Foscolo Benedetto, "is a secular and earthly synthesis to be placed next to the two famous syntheses in which the theological and philosophical Middle Ages were summed up, the Summa of Saint Thomas Aquinas and the Divine Comedy of Dante Alighieri." In the following, we will not have the opportunity to deal with Aquinas's Summa as much as we would like, but *Il Milione* and the *Commedia* will keep us company on a great journey along the roads of the world and those of the mind, in the infinite power granted to us by our imagination.

The Giant Crystal

"The best book written by men," says Borges, as if to suggest that a better text could only be written by the gods. The Divine Comedy is a book for crystallographers, because it is a single, large crystal with 14,233 faces. A crystal that reflects and refracts light as in a gigantic kaleidoscope, in which the ray of light emerging from one edge re-enters another, creating from nothing a grandiose and paradoxical hologram of the harmonia mundi of the symbolic, sacred and scientific thought of our distant Middle Ages.

There is a noble branch of mathematics, the group theory, founded by the unfortunate Évariste Galois, that is responsible for the study of all symmetries in nature: in addition to the secrets of elliptic curves or multidimensional spaces, and besides finding an answer to the thousand questions posed by the forces that shape our Universe, this theory should also unveil all of the scores, the permutations, the forms of energy hidden in the back of the incomparable ocean of words embraced by the Comedy. Behind these scores, there is the iron logic of geometry, of the celestial arcs of astronomy, accompanied by the breath of history and the music of poetry. There is the musicality of the verse, which cleverly appeals to the memory, a facility that, in medieval readers, was infinitely more refined than ours, due to the endemic absence of written texts. In fact, the *Commedia* was not even finished when it began to circulate orally in the villages and courts of Italy: rather than reading it, people knew how to recite it by heart. The force of the endecasyllables and the rhymes of the tercets guided the language, forced the breath. Verses such as.

'Charon,' my leader, 'don't torment yourself.
For this is willed where all is possible
that is willed there. And so demand no more.'

Inferno, III

immediately stood out for the rhythm of their rhymes and the eloquence of their words.

The Theory of Beauty

"Thanks to numbers everything becomes beautiful", Pythagoras said. And it was the Pythagorean school that was the first to transfer the ideal of measurement, associated with numbers, to the very concept of beauty. Beauty, for the Pythagoreans, coincides with the order of things, which can be found not only in the world of ideas, but also in the world around us, for example in the curve of life, visible in the spiral of the Nautilus shell, in the regularity of the faces of a crystal or in the hexagonal shapes of snowflakes. "The proportion of the parts" is the guiding principle underlying the very idea of harmony and can manifest itself in both the interweaving of music and the most daring architectural forms. In this formulation of aesthetics, natural beauty and artistic beauty come to coincide. This tradition has greatly influenced our entire culture, even in the field of science: Galileo, Kepler and Copernicus himself would be inconceivable without a conception of the cosmos as the place of beauty itself.

The idea of order of the ancient Greeks played a fundamental role for all art in the Middle Ages and the Renaissance, and the echo is in the verses of our Dante:

The things all
have order between them, and this is form
That the universe is like unto God.

Paradise, I

In the field of mathematics, considered by the great Carl Friedrich Gauss the "queen of the sciences", the theory of beauty came to life with a young

> Parisian born in 1811 and who died very young in a mysterious duel in 1832: Évariste Galois. Galois spent the evening before the duel correcting the manuscript that laid the foundations of what would later be called group theory, one of the most fascinating and pervasive topics in science, called upon to identify the concept of beauty and symmetry - wherever it may be hidden in nature - behind a geometric construction, the behaviour of an elementary particle or the proof of Fermat's theorem. Group theory also tries to classify all the possible ways in which symmetries can occur, because this helps us to better understand the laws of crystallography even for objects that our eyes cannot see. The largest exceptional symmetry has the Dantean name of Monster Group and lives in a space of 196 883 dimensions. In spite of its name, this is a "mathematical crystal" of enormous structure and exquisite beauty and complexity which, with each new analysis, reveals ever new and fascinating details, revealing coincidences on the deep physical nature of the Universe. We like to think of the Divine Comedy as an enormous, fascinating crystal that continues to reveal surprising details about man, the most mysterious being in all of creation.

Science teaches us that, in order to understand a phenomenon, we must first look at it from a distance, trying to grasp its essential features. In the case of the Divine Comedy, this is an almost obligatory exercise in order not to become prisoners of the crowd of unlikely characters that populate its verses and the thousands of details connected with them. In the following, we will therefore take a quick overview, necessarily partial and superficial, of this gigantic work whose main axes are wisdom and history, letting ourselves be guided in this summary by suggestions, but also by simple curiosities, almost as if pretending to be a foreign reader approaching such a unique book for the first time. We will then enter into the lattice of this magnificent crystal in the next chapters, when our sight will be weaned and adapted to both light and darkness.

The first thing that the reader realizes when reading the Comedy is that we are talking about a dream, a great hallucination, as can be seen from the systematic violation of all space–time logic. It is the story of a journey made by a pilgrim—the Poet himself—into the three realms of the Underworld, accompanied by three exceptional guides: first, Virgil, the great pagan poet, "wise and noble, knowing all", then, Beatrice, the young Florentine noblewoman whom Dante had loved with infinite trepidation:

seen through a veil, pure white, and olive-crowned,
a lady now appeared to me. Her robe was green,
her dress the colour of a living flame.

Purgatorio, XXX

and finally, Saint Bernard, the great contemplative mystic:

> *Now Bernard, smiling, made a sign to me*
> *that I look up. Already, though, I was,*
> *by my own will, as he desired I be.*

Paradiso, XXXIII

In this journey, the pilgrim crosses Hell, the world of the damned; then, along the slopes of the mountain of Purgatory, he meets the penitents who are waiting to expiate their punishment in order to reach Paradise, the place where the truly blessed stand in adoration of God.

The journey to Hell covers all possible declinations of human evil: lustful, greedy, avaricious and prodigal; heretics and murderers, sodomites and suicides; deceivers of all kinds and traitors of all kinds; it is, at the same time, a long, painful and articulated journey of formation for the pilgrim and his wise master Virgil. From the frightening "dark wood" to the encounter with Lucifer himself, the journey into the realm of the dead returns the allegory of the whole culture of the Christian Middle Ages to us in verse form, and the Poet's path of redemption ends up being an opportunity to speak not only of the world of the dead, but also of the world of the living, and of all of us.

The ascent of the mountain of Purgatory—step by step, leap by leap—underlines how only through expiation, reflection and repentance can we aspire to redemption. At the entrance to Purgatory, he is greeted by Cato Uticense in the morning beach air:

> *Soft hues of sapphire from the orient,*

Purgatorio, I

Then, it's just a long climb, until he meets Beatrice, who is waiting for him in the "gentle breeze, unchanging in itself" of the Earthly Paradise on the top of the mountain. Between the initial meeting with Cato and the final one with Beatrice, Dante gets to know all the shades of the seven deadly sins—pride, envy, wrath, sloth, avarice, gluttony and lust—directly from the voices of those who were stained by them in life. Like the medieval pilgrims on their way to Santiago de Compostela, all these souls in pain are on their way to the top of that mountain, where they will be immersed in the waters of the River Letè, which cancels the memory of their sins, and of the Eunoè, which strengthens the memory of the good done in earthly life.

Finally, with his ascent into the immaterial ethereal world of the heavens of Paradise, Dante knows eternal bliss, and as he ascends to the Empyrean and the Rose of the Blessed, around him increases the light and Beatrice's smile becomes more and more dazzling, until the moment she reaches God,

love that moves the sun and other stars.

Paradiso, XXXIII

This is a terribly sketchy description of Dante's masterpiece. Thus, it is worth accompanying it with a series of comments linked to its genesis and medieval traditions. In fact, if the Comedy remains the summa of all imaginary journeys beyond the world, with the majesty and fantastic rigor of its verses, there are, however, stories and legends that have their roots in the dawn of civilization and that prepared the ground for it. It is Dante himself who recognizes this, taking care to recall, at the very beginning of his work—in Canto II of the *Inferno*—that, before him, both Aeneas and Saint Paul had traveled to and returned from the afterlife.

As far as Aeneas is concerned, the reference is almost obligatory, given the guide upon whom Dante relies for that curious journey. In fact, it was Virgil himself who recounted how Aeneas, one night, while he was in Sicily, had seen the shadow of his father Anchises in a dream and how his old father had invited him to stop at Cumae in order to descend into the Underworld and know his future, entrusting himself to the Cumaean Sibyl, Apollo's priestess. The account of that journey is what animates the most evocative pages of the Aeneid: after having crossed a swarm of monstrous figures at the entrance to the Underworld, Aeneas arrived on the bank of the river Acheronte, where Charon was waiting to ferry him to the endless plain where the souls who had died before their time were resting; he then passed by the triple wall of Tartarus, the horrendous infernal city, and after having met Cerberus, a terrifying three-headed dog, he entered the Fields of Wailing, the place of suicides. And here, amid that multitude of unfortunate people, Virgil tells us that he saw Dido, the queen of Carthage, gaining confirmation of what he already suspected, that is, that Dido had committed suicide for him: in the face of his words of justification ("fate", "destiny", "the great mission that awaits me"), Dido opposed him with an icy muteness. After this sad encounter, Aeneas was finally able to reunite with his old father in the immense meadows of the Champs Elysees, radiating with a purple light. And, after exchanging a triple embrace, Anchises, having climbed a hill, pointed out to the Trojan hero the souls of the generations that would make the future of Rome glorious.

Characters, monsters and places such as Dido, Cerberus and Acheron pass from the Aeneid to the Comedy and reappear unchanged from the shadows of the past. It is no surprise, therefore, that it would be Virgil, an expert connoisseur of the otherworldly world, who would accompany Dante on the first part of his journey: it is a way of openly underlining the source of inspiration of various scenes and figures in Dante's work, that is, of not concealing the common root that binds the myths of all times.

As for Saint Paul, the main missionary of the Gospel among the pagans and Romans, the afterlife journey of which he was the protagonist is recounted in an apocryphal text probably dating from the fifth century AD, a text that elaborates and amplifies a brief mention of this event in Paul's letter to the Corinthians. The story is that of Paul's abduction by an angel sent by God to show him the state of souls after death; after being shown all of the righteous folk, Saint Paul is led to the promised land and to the shore of an enchanting sea, where the City of God is located: this is the place set aside for prophets, saints, patriarchs and all of the blessed. A river of fire separates him from Hell, and once he has crossed it, Paul sees the place where the lost souls are confined, arranged in circles according to the transgression committed. His otherworldly journey finally ends at the threshold of Earthly Paradise.

There is, however, a third astonishing source of inspiration for Dante's great poem, and that is the arabesque world of Islam, which goes far beyond the influences of Averroes,

Averroes, too, who made the Commentary.

Inferno, IV

of Avicenna or Saladin, all cited in the Comedy with great respect among the eminent scholars who dwell in the "noble castle", as we shall see. It therefore seems that, in addition to Saint Thomas, Aristotle, Ovid or Cicero, in Dante's imaginary, boundless, but unknown library, there was also room for texts from the Near East.

The main proof is in the *Book of Muhammad's Ladder*, an ancient, mysterious and anonymous eschatological text of Arab-Spanish origin that, developing a famous Koranic verse, tells the story of Muhammad's nocturnal journey to the otherworlds, starting from Jerusalem. The analogies with the Comedy are surprising: guided by the angel Gabriel, Muhammad ascends to Paradise by means of a shining staircase, crosses eight circles of heaven, meets eight prophets, enjoys the delights of nature and love of the place, then arrives at the level of God, who entrusts him with the Koran and speaks to him of the precepts of daily prayers and fasting; he then visits the seven infernal lands, immersed in the gloomy atmosphere of the torments and cries of the damned, and finally returns to earth to reveal his vision to the inhabitants of Mecca.

The world of Islam's influence on Dante does not stop here, however, as originally shown by Miguel Asín Palacios, a Spanish historian and Arabist, who, in a disruptive work written in 1919, declared that he had traced the events and models to which Dante would have turned to draw real inspiration for his Comedy. Palacios' text was destined to shake the consolidated certainties of traditional Dante writers and to fuel a fervent debate, which, over the years, has involved not only the esoteric voice of René Guénon, French writer, philosopher and occultist, but also the voice of the well-known Italian philologist and writer Maria Corti.

All of these scholars have repeatedly pointed out the myriad connections with Arab and Islamic traditions scattered throughout the *Commedia*. The first example concerns a Muslim legend in which a wolf and a lion block the pilgrim's way, in the same way that the lion, the lynx and the she-wolf prevent Dante from climbing the sunlit hill and cause him to retreat; also worthy of note are the similar roles played by the Archangel Gabriel and Virgil in these two respective afterlife journeys of Mohammed and Dante.

The *Inferno*, too, with its deafening voices of pain and its gusts of fire, has important traits in common with other narratives: in both the *Inferno* and the *Book of Muhammad Ladder*, a gigantic funnel opens up beneath the feet of Jerusalem and leads to the centre of the Earth in a decline of plateaus and circular staircases, with sinners arranged along circles that grow deeper and deeper as their guilt worsens. The punishment of thieves tormented by snakes is described by Dante in the same words as the anonymous Arab author of the *Book of Muhammad's Ladder*. Dante's triple ablution upon leaving Malebolge—a cursed place in the deepest part of Hell where seducers and pimps, thieves and fraudsters are punished—is the same triple ablution that purifies souls in Islamic tales.

The design and edging of the heavens have the same patterns and punctuation: in both the *Comedy* and the *Book of Muhammad's Ladder*, the souls of the blessed are arranged along nine heavens and the final apotheosis is accomplished in the presence of a Creator, a source of blinding light, surrounded by festive angelic hosts.

"A clue is a clue, two clues are a coincidence, but three clues make a proof," said Agatha Christie, the undisputed queen of the mystery novel. So, the question now is no longer whether Dante drew on Islamic sources, but rather how he did so.

The most suggestive thesis, credited to Maria Corti, involves an enlightened sovereign, Alfonso X of Castile, a lively coterie of scholars of all races and traditions, the court of Toledo of the twelfth century, and a great man of culture in exile, Brunetto Latini. The latter, a Florentine politician of tremendous culture upon whom we will dwell at greater length later, was sent by his fellow citizens to Toledo to ask Alfonso X for help on behalf of the Guelph side to which he belonged. But the news of the disastrous defeat suffered by the Guelphs in the battle of Montaperti on September 4, 1260, took him by surprise while he was at the court of the Castilian sovereign, with whom he had to forcibly prolong his stay so as not to incur the reprisals that awaited him if he returned to Florence.

The most accredited hypothesis, therefore, is that Brunetto read the Book of *Muhammad's Ladder* at the court of Toledo, where the volume had been translated from Arabic into Castilian by the Jew Abraham Alfaquin, Alfonso X's doctor, and, again at the same school in Toledo, had then been translated, in 1264, from Castilian into Latin and Old French by Bonaventura da Siena, who was also an exile and a guest at the same court. And so, Brunetto Latini, Dante's future teacher, is the main suspect in this Islamic impression on our supreme poet. Dante's omnivorous curiosity would do the rest.

Acknowledging these influences obviously takes nothing away from Dante's greatness, much to the annoyance of Palacios' ancient opponents. How many great authors, in fact, have turned a deaf ear to previous literary traditions in their attempts to conceive an absolutely original work? Let us think, for example, of the magical lightness of Ariosto's Orlando Furioso or, in more recent times, of the enigmatic origin asserted by Borges for many of the stories in Fictions, inspired, according to him, by real or imaginary texts.

The Enigma of the Lost Manuscript

Let's open a parenthesis on a real bibliographical mystery with something incredible to offer: to this day, we know of no documented autograph by Dante: no letter, no signature affixed to a document, or page written in his own hand in one of his many works. Nothing at all, a paradoxical thing, especially considering that the Divine Comedy is not only one of the most widespread works after the Bible, but also among the most voluminous! It is unknown, in fact, what happened to the original manuscripts of the poem: do they still exist? Are they hidden somewhere? Many scholars are convinced of this, including on the basis of the sheer volume of the work, as we mentioned.

Moreover, it is very probable that there were at least two copies autographed by Dante: suffice to think of Dante's famous epistle to Cangrande of Verona, his protector, in which he writes of wishing to dedicate *Paradise* to him. It is likely, in fact, that he accompanied the epistle with a copy of the poem, given that he goes on at length about its contents and allegorical style: what sense would there have been in writing to Cangrande about a text in such great detail and dedicating it to him if the work in question had not been in his actual possession? A copy of the Comedy, therefore, or at least parts of it, must surely have arrived in Verona, and from there ended up in who knows which corner of the palace of Santa Maria Antica, where Cangrande resided, or perhaps it was given as a gift to some of his visiting guests.

It is clear that one of the problems at the basis of the disappearance of Dante's manuscripts and the *Commedia* lies in the Poet's continuous movements during his life in exile: he began writing the first Cantos of the *Inferno*, perhaps, in 1304, or more probably in 1306 (based on texts written earlier in Florence when he was not yet in exile), and finished the last Cantos of *Paradise* in Ravenna in 1321, the year of his death. In the years 1314–1315, the *Inferno* was made public in Verona, and the *Purgatorio* the following year, while the *Paradiso* would be released by the Poet's sons, Pietro and Jacopo, after his death.

In this regard, there is the singular story of the discovery of the last thirteen Cantos of *Paradise*, which had proved impossible to find at the time of Dante's passing, but then miraculously reappeared thanks to a dream that was had by his son Jacopo.

Giovanni Boccaccio, a great admirer of Dante and author of the *Trattatello in laude di Dante*, wrote about all this. In it, Boccaccio recounts that Dante, having finished *Paradise*, decided to hide his last Canti. When he died, Jacopo and Pietro put their father's papers in order and made the bitter discovery that, in the last Canto, thirteen Canti were missing. They searched for a long time, but without success. A year went by, until, one night, Dante appeared to Jacopo in a dream and showed him the place where he had hidden the missing Canti: they had been placed in a crack in the wall of the house in Ravenna where the Poet was living at the time of his death, a hole hidden from view by a curtain.

Equally singular is the story of the oldest known transcription of the verses of the Comedy: it dates back to 1317, and is credited to the notary ser Tieri degli Useppi of San Gimignano, who wrote some verses of the III Canto of the *Inferno* on the cover of a register of criminal acts, currently kept in the State Archive of Bologna. Other notarial acts of the time involving transcriptions of certain verses from *Purgatory* are also deposited in the same Bologna Archive: the prosaic reason for the transcription lies not so much in those legal officials' love of poetry as in their need to fill in the blank sections in the pages of their documents to prevent additions or changes by third parties!

Between 1322 and 1330, in Rome, a rabbi transcribed some tercets of the *Paradise* in Hebrew, but we have to go back to 1336 to find a full transcription of the Comedy, made in Genoa by the copyist Antonio da Fermo: that parchment manuscript is now preserved in Piacenza at the Biblioteca comunale Passerini Landi and, for this reason, is called the Landiano manuscript.

The following year, it was Francesco di ser Nardo da Barberino who copied the manuscript, which, in this case, is called the Trivulziano, because it is kept in the Trivulziana Library in Milan. In 1347, Francesco da Barberino directed a workshop of copyists that, in a short time, produced about a hundred copies of the *Commedia*.

The real editorial promotion of the work, however, came from Boccaccio, who, in Florence, between 1374 and 1375, read the poem publicly in the church of Santo Stefano in Badia and made three copies himself, one of which was given as a gift to Francesco Petrarca.

Where could the originals of the *Commedia* be? The most accredited hypothesis is that they are in Verona, buried in some library or on the shelves of an archive. Another suspected location is Pomposa Abbey, along the Via Romea, where Dante stayed many times during his diplomatic missions and whose monks were among the first to produce copies of the poem. And, finally, some consider the Vatican Library, to which the volume may have been donated by Dante's nun daughter, Antonia. But if it was a gift from Dante's daughter to the pope, the manuscript could also be in Avignon, because, in the years in question, the seat of the popes had been moved to France and the pontiffs no longer stayed in Rome. The mystery continues.

The Disease of Writing

Before he came to compose the immortal masterpiece that we know, Dante had gone through several stages of writing, as if in a constant apprenticeship, in order to be able, when the time came, to scale the lofty heights of literature. He was not the first, and obviously not the only one, to do this: after him, Proust, for example, in advance of the monumental cathedral of *In Search of Lost Time*, had tried his hand at writing pages and pages of the novel *Jean Santeuil*, eventually left unfinished; Marie-Henri Beyle, better known as Stendhal, before his masterpieces *The Charterhouse of Parma* and *The Red and the Black*, had honed his pen composing a series of essays on Gioachino Rossini, Jean Racine and William Shakespeare, as well as on his beloved Italian cities of Rome, Naples and Florence.

In Dante's case, his first literary steps were taken in the company of the poets of the "stil novo," following a crucial encounter with Guido Cavalcanti and the cenacle of other refined poets, such as Lapo Gianni and Dino Frescobaldi, and study of the verses of Guido Guinizelli and the courtly poetry of Guittone d'Arezzo. These were the youthful years in which Dante was enraptured by the enchantment of escape from the reality of "boring people," the years of poetic training, of learning to rhyme his thoughts, to explore the sound and meaning of the new words of the vulgar tongue:

Guido, I would wish that thou and Lapo and I
were taken by enchantment
And put into a pot, which at every wind
By sea went to your will and mine.

Rime, LII

Dante soon distanced himself from this group, however, because, in the meantime, he had developed different ideas about women, the symbolic value of love and the importance of philosophy. The new phase is described in poetry and prose in the *Vita nuova*, where we find the upheaval of perspectives brought about by the apparition of Beatrice, a story that, beyond the glances, the palpitations and shortness of breath at the sight of the beloved, is immediately projected into a space and time that has no connection with reality.

The encounter, at nine years old, of the Poet with the "kind one" leads him to touch "all the terms of beatitude," but, once the noble maiden dies, the celebration of her beauty and her love turns into an embrace of the splendour of celestial glory, because it is ineluctably linked to the anxious desire for knowledge and the search for divine contemplation. As he affirmed at the end of the *Vita nuova*, he would no longer write about Beatrice until he could "more worthily deal with her": we are in 1295, and almost ten years would pass before the "kindest" would appear again, in the verses of the Comedy, becoming the cornerstone of his majestic building.

The Contest with Forese

Paraphrasing Joyce, Dante's portrait of the artist as a young man also passes through the notes of the comic, the scurrilous, the taste for heavy jokes and offence. The most famous example is an exchange of sonnets he made with his friend Forese Donati, known as Bicci, around 1290. Forese, who died in 1296, was a member of an important Florentine family. He was the brother of Corso and Piccarda Donati and therefore a distant relative of Gemma, Dante's wife.

> This lively tussle is made up of six sonnets, three by Dante and three by Forese, and in each of them references are made to the private lives and families of the contenders and no insulting accusations are spared. In the fourteenth century these verbal clashes were quite common in the world of jesters and represented the very soul of the cabaret of the time, where insults and jokes were functional to seek applause or provoke the roar of laughter from the audience. Dante opens the hostilities, making heavy irony about his friend's virility and his infamous poverty. Forese retorts that he had met his father's ghost and that he wasn't doing so well, seeing that he had been buried in a mass grave in deconsecrated ground, a fate that usually befell only heretics, usurers and even those who didn't have the money for a tomb. Dante changes battlefield and accuses Forese of being a glutton, while the latter retorts that Dante eats at other people's expense. Tones rise, tempers flare: Dante insinuates that Forese is not his father's son and that he steals to satisfy his gluttonous sins; Forese, at this point, retorts that it is better to be a bastard son than Alighiero's son, given Dante's cowardice in carrying out a vendetta on behalf of his father.
>
> The accusations could go on indefinitely: infidelity, cowardice, thievery have always been gold mines for insult and invective, for a repartee that leaves as the only winner the most skilled with the tongue and words, used as darts.

In the meantime, Dante experienced the bitterness of exile, the jealousy, envy and resentment of his fellow citizens, but the uncertainty of those years did not prevent him from conceiving ever more ambitious literary projects, to which he would submit his extraordinary intelligence, his superb wisdom: between 1304 and 1307, he launched into the enterprise of compiling the entirety of human knowledge in a universal encyclopaedia, a book that would have made Borges happy, because it would have housed the names of all things inside of it, finding a word for every idea and an idea for every word. Dante wanted to prepare a banquet of wisdom and, for this text, he chose the name *Convivio*. The work also had a pedagogical intent, an opening towards all those who might wish to dedicate themselves to study, but had not been able to do so due to "family and civil cares." The language, therefore, could only be the vernacular, the language of the "non-literate."

This ambitious editorial plan dedicated to the praise of wisdom, however, got bogged down after a few steps: in the initial phase, fifteen treatises were conceived; ultimately only four were actually finished. The problem was that Dante, while doing one thing, was always thinking about a hundred others, and it was not unusual for him to suddenly throw himself into another venture.

One of the issues closest to his heart in those years was, in fact, the promotion of the vernacular language, a lexicon whose enormous expressive potential he had glimpsed, as if it were a new musical instrument with which to compose polyphonic songs never heard before. So, he wanted to dictate the rules immediately, to set down in print the methods to be used in that new art of writing in the vernacular. In addition to strengthening and reinforcing the ties that already existed, it was also a matter of inventing this language, of enjoying the sounds of its words, of playing with their ambiguities or sharpening the rigour of their definitions. Like a wizard genius, he invented a beautiful image for this coveted idiom, the "perfumed panther."

In fact, he had an even greater ambition: to convince people on the importance of the vernacular. But in what language do the learned communicate? In Latin, of course. And so, you get the exquisite paradox: Dante writes a treatise in Latin to promote the qualities of the vernacular! He calls it *De vulgari eloquentia*, a name that logically twists upon itself, almost like the aporia of the Cretan who claims that all the inhabitants of Crete lie.

All of these reflections on language will be very useful to him later on, particularly when he has to try to make intelligible the secret of the divine vision that presents itself to him at the end of the great otherworldly journey: Indeed, it is in *Paradise* that he will treasure the malleability of the language to which he has entrusted himself, using the very metamorphosis of that otherworldly place to coin entirely new phrases and verbs; a veritable festival of neologisms, such as, for example, "indovarsi," which expresses the concept of "taking place" in the verses "I willed myself to see what fit there was, image to circle, how this all in-where'd" to try to grasp the mysterious union of Christ's two natures.

However, the grandiose editorial project linked to *De vulgari eloquentia* foundered like that of the *Convivio*. What remains of the work, however, gives us a glimpse of the virtuosity and genius that had driven it in the first place: the main question of this book revolves around the question of the search for an "illustrious vernacular," that new ideal language capable of absorbing the best of the fourteen dialects identified at the time by Dante and of electing itself to the level of universal idiom, a sort of versatile Esperanto in which to express the chronicle of history, philosophical disquisitions, literary contests, poetry, wise treatises, and emotions.

The apprentice writer's splendid career ended with another stupendous utopia: a treatise, originally conceived in three volumes, on the role of the Empire and its relations with the Church, a dramatically topical theme in the fourteenth century, as society had been torn apart by the struggles between Guelphs and Ghibellines, between those who vehemently maintained that the moral and political guidance of peoples was the responsibility of the pope and those who asserted just as vehemently that it was the responsibility of the emperor.

The title in this case is simply *Monarchia*, in which Dante astutely seeks to find a third way to the frontal opposition of the two parties, manifested as the theory of the two Suns: if the power of the Emperor must be autonomous, so must that of the Church, because these powers descend directly from God and have different aims: for the Empire, happiness on earth, for the Church, happiness in heaven.

His destiny as a writer took a fortunately different turn at this point, no longer dealing in fragments of speeches on love, approximate texts on wisdom, philosophy or theology, utopian proposals on politics or paradoxical presentations on the virtues of the vernacular: what awaited him was the greatest, most colossal, majestically political, philosophical, theological and scientific fresco that could ever be conceived, written directly in the vernacular. A text moved by the strings of love: the Comedy. As the Russian poet Osip Mandel'štam wrote:

> If the halls of the Hermitage suddenly went mad, if the paintings of all the schools and all the great painters suddenly came off the nails, penetrated each other, mixed with each other and filled the closed air with Futurist screams and furious chromatic excitement, the result would be something like Dante's Comedy.

It's now time for us to learn more.

In the Darkness of Hell

Sighing, sobbing, moans and plaintive wailing
all echoed here through air where no star shone,
and I, as this began, began to weep.

Inferno, III

On June 21, 1527, shortly before his death, Niccolò Machiavelli had a dream, and took the trouble to tell his friends gathered at his bedside. In the dream, he had seen a procession of men, with emaciated and tired faces, dressed miserably. When he had asked them who they were, he had been told that they were the saints and the blessed, marching towards Paradise. Turning his head, he had then seen a host of other men of solemn appearance and noble countenance, discussing, in grave voices, important aspects of wisdom. He had recognized among them the philosophers Plato and Aristotle, the historians Tacitus and Plutarch, the poets Horace and Lucan. He also asked them who they were and where they were going. He was given the answer, "We are the damned of Hell." Then, in a feeble voice, but accompanied by that ineffable smile with which he had always looked at life and men, Machiavelli told his friends that he would much rather have gone to Hell, to be able to talk politics, philosophy or history with those great men, than to sit among the saints, blessed by the unbearable joy of boredom.

Machiavelli's ironic reversal of the scale of values is the same as that shared by many readers of the *Commedia*: there are many, in fact, who still think that the horrid but vivid scenes of Hell are far more exciting than the learned theological disquisitions of *Paradise*.

For Dante, the pilgrimage to the realms of the Underworld must necessarily start from the squalid spectacle of a thousand sinners, the scene of the world's wicked corruption, a necessary stage before reaching the glory of Paradise, the redemption, and redemption not only of himself, but of all humanity. Hell (which speaks to us of "unpleasant things") and Paradise (which concerns "light") will therefore necessarily have the same length within the gigantic crystal of the Comedy. Purgatory too, of course, but this is a separate issue that we will deal with more fully later. Dante's entire journey touches on all of the great themes of humanity and literature: desire, the passing of time, the power of memory, revenge, enchantment, love, hatred, loyalty and betrayal, destruction, the bitter chalice of exile and the sweetness of finally returning home.

The first thing that strikes the reader in the Comedy is the fairy-tale tone with which the story opens and, at the same time, the curious mixture between the narrator and the protagonist of this strange adventure. As if that were not enough, we must then add a third dimension, which is the biographical one. A poem therefore one and three, with temporal planes that cross and intersect, in an eternal cross-reference between dream and reality.

While the protagonist's otherworldly journey begins and ends within the period of a week, the writing process took the narrator about a dozen years, from 1307 to 1321, the year of his death: a crucial period in which the flesh-and-blood Poet experienced the bitterness of exile, wandering from court to court begging for hospitality and protection, pursued by condemnation for "barter, fraud, falsehood, malice, iniquitous extortion practices, illicit proceeds, pederasty," for which he was subjected to "a 5000 florins fine, perpetual banishment from public office, perpetual exile (in absentia), and if he is caught, at the stake, so that he dies," as the sentence issued in 1302 by the city court of Florence reads.

That otherworldly journey opens with the protagonist lost in a dark forest, from which he tries to extract himself, after initial bewilderment, by climbing the crags of a "delightful" hill that shines in the sun. It is a mise-en-scène of sin and evil in whose plots the protagonist is lost and, with him, the whole of Christianity. Possible redemption, however, seems within reach, attainable if one counts on reason and virtue. But the road is immediately blocked by three beasts, whose names all begin with L[1] (the lion, the lynx, the she-wolf), in kinship with the L of the perfidious Lucifer, supreme symbol of evil, with whom they are united by lust, pride and greed. These animals come directly out of the bestiary of the Middle Ages, that immense herd of real and fantastic beings of which the illuminated codes present in the convents and capitals of the Romanesque churches were full. The nature of these animals that block the Poet's way to the hill announces the division into three great families of sins into which Hell will be divided: incontinence, bestiality and malice.

To save the lost Dante (the name of the protagonist, as we will learn only in the fiftieth verse of Canto XXX of the *Purgatory*), a shadow appears here.

to him I screamed my 'Miserere': 'save me,
whatever—shadow or truly man—you be.'

Inferno, I

It is Virgil, the great Latin poet, who, in the Middle Ages, was the very symbol of boundless knowledge and who, within the Comedy, anticipates all of the other father figures that the protagonist will later meet during his journey, such as the rhetorician Brunetto Latini, the musician Casella, and the adventurous Ulysses.

The shadow reveals to him that, in order to save himself, he must follow another path and that, paradoxically, the way to his salvation does not consist in going upwards but rather downwards! In fact, he will have to descend, like a speleologist of the soul, into the bowels of the Earth, into the dark and frightening chasm of Hell, where he will hear the desperate cries of the damned, only later to be able to climb up the hill of Purgatory, where he will see the souls of those happy to suffer because, after their suffering, beatitude awaits them. In Paradise, however, he will be guided by a soul more worthy than Virgil, a soul who, as a pagan, is denied the vision of God and access to the city of the saints. The soul in question is the "most gentle" Beatrice: with the Comedy, Dante can finally honour his promise to write more worthily than her.

---

[1] In Italian "lonza, leone, lince".

> **The Hell of Primo Levi**
>
> Primo Levi, a chemist of Jewish origin, was arrested in 1943 by the Fascist militia and interned in the Auschwitz extermination camp. The abyss of evil into which man had sunk opened wide before his eyes, that indecipherable hell apparently without boundaries or shame. The dramatic experience he lived in that concentration camp was the theme of If This is a Man, a novel in which, in order to find the most suitable words to describe all the atrocities of which he was a victim, he turned to Dante's Inferno. Every reference in the great poem seemed to have a parallel in that Anus Mundi: the door of Hell and the inscription on it were superimposed by the mocking inscription "Arbeit Macht Frei" (work makes you free) on the gate of the camp; in place of the horrible Charon here was a German soldier taking away gold, money and watches to those who got off the armored cars; the infernal judge Minos was associated with the SS officer who decided who was able to work and who should be immediately sent to the gas chambers, and so on.
>
> With the dry and rigorous style of a scientist but with thoughts of the Comedy, Primo Levi reconstructed the entire journey to the end of the night experienced in that dark corner of Poland, that infinite degradation towards "a blurred, frenetic film, full of noise and fury and without meaning: a hustle and bustle of nameless and faceless characters drowned in a continuous deafening background noise, on which, however, the human word did not emerge.
>
> When Levi meets Jean, a young Alsatian student who wanted to learn Italian, his thoughts fly to Canto XXVI of the Commedia and to Ulysses. Levi launches into the crazy enterprise of telling Jean who Dante is, how the Commedia is organized and what the Inferno is like. The pivotal verse of Ulysses' entire experience—"You were not made to live like brutes, but to follow virtue and knowledge"—dazzles them as if it were the voice of God, commanding them never to lose their dignity. "For a moment, I forgot who I am and where I am," wrote Primo Levi. He was one of 20 survivors of the 650 Italian Jews who arrived at the camp.
>
> Years later he wrote that if "there is Auschwitz there can be no God. I cannot find a solution to the dilemma. I look for it but I don't find it." For him, the Comedy unfortunately stopped only at Hell.

Virgil then launches into an enigmatic prophecy addressed to the ravenous beast that has blocked Dante's way, the she-wolf: he declares that the time will come when a hunting dog, with the enigmatic name of Veltro, will appear on the scene and make her die painfully. And, to further muddy the waters and to meld dream and reality, myth and history, the narrating Dante makes Virgil say that this Veltro will also be the salvation of Italy, the country for which both the virgin Camilla and Euryalus, Turno and Niso died, uniting, in one fell swoop, Italic and Trojan characters, real and imaginary figures.

A strong fellowship is immediately born between the two poets: Dante's immense admiration for that spiritual father sent to him directly by Beatrice combines with Virgil's tenderness for that lost soul. Osip Mandel'štam cannot help but observe that, throughout the entire Comedy, "Dante does not know how to behave, where to put his feet, he does not know what to say, nor how to make a bow": in the journey through Hell and Purgatory, it is Virgil who attends to the protagonist's etiquette, leaving this thankless task to Beatrice once they reach the threshold of Paradise.

The journey begins on the night of Good Friday in 1300, when Dante and his guide pass through the terrible gate of Hell.

'Through me you go to the grief-wracked city.
Through me to everlasting pain you go.
Through me you go and pass among lost souls.
[...].
Surrender as you enter every hope you have.'

Inferno, III

and they enter the chasm that opens frighteningly at the foot of Jerusalem. They will emerge exactly three days later, in a parallel with the resurrection of Christ after his crucifixion, one of the thousands of symbolic references with which the Comedy is full.

Allegory is one of the rhetorical figures par excellence for the men of the Middle Ages, given their propensity to interpret earthly reality as a discourse that always refers to another reality, one otherworldly and divine. Everything is, therefore, a symbol, as witnessed by the lapidaries, herbariums and bestiaries of the time, in which moral meanings, virtues or negative influences are dealt with through attribution to stones, herbs or beasts. The Comedy is therefore a gigantic allegory, in which the historical episodes and characters have both real and symbolic value. We will see later the great symbolic value assigned by Dante to numbers, and we will better understand, for example, the frequency with which the number three appears in the work: three Canticles, for example, three beasts, three faces of Lucifer.

Incidentally, Dante traces the geological origin of the infernal chasm back to Lucifer, who, having rebelled against God, was hurled from Paradise to earth: horrified by the contact with this unclean being, the emerged lands of the northern hemisphere retreated and went on to form the mountain of Purgatory in the southern hemisphere, the only point not submerged by water in that part of the globe.

Having passed through the gate of Hell, Dante and Virgil arrive on the banks of the Acheronte, the first of the infernal rivers, created, like the other two, the Styx and the Flegetonte, by the tears of the great Veglio of Crete: according to the legend, this was a huge statue of an old man enclosed in a cave within Mount Ida, on the island of Crete, with a head of gold, a chest and arms of silver, a trunk of copper, and legs and a left foot of iron, while the right foot, upon which rested the weight of all of that metal, was made of vile clay. The tears gushed from the wounds that filled the body of the Veglio and, gathering at his feet and flowing between the rocks, gave rise to an underground river that, in its path in the bowels of hell, took, in turn, the names of Acheron, Styx and Phlegeton, ultimately falling into the lowest chasm of Hell and forming the frozen lake Cocytus.

On the banks of the Acheronte, the first close encounter with a real infernal character occurs, the ferryman of the lost souls of the world, "Caron dimonio, con occhi di bragia." In the darkness and amid the screams of the damned with which the cavern echoes, here comes that "old man, white by ancient hair," who does not hesitate to beat with his oar those who lie down among those wretched. The flash of vermilion light that rips through that apocalyptic scene anticipates the arrival of an earthquake that shakes the earth and causes Dante to faint, slumping to the ground as a "dead body falls." When he wakes up, he is across the river, a symbolBut what does the frightening abyss of Hell look like? The funnel of Hell is divided into nine concentric circles, whose circumference gets smaller and smaller as you go down, until you meet Lucifer. The lower you go, the farther you are from God and the greater the gravity of the sin punished. The infernal judge is Minos, the animal-like figure who "snarls" and "examines faults," and who indicates to each damned person the circle to which they are destined by wrapping his tail around his enormous body a corresponding number of times.

From the second to the fifth circle are the incontinent, that is, the lustful, greedy, avaricious and prodigal, wrathful and lazy, all those who have not been able to resist the primal instincts and drives of their bodies. The first circle is, in a way, a very special circle: besides the children who died before baptism, there are the great wise men of the past—poets, philosophers, men of science, princes or pagan heroes—who live in a "noble castle" surrounded by seven walls and defended by "a beautiful little river." It was within the walls of this castle that Virgil himself stood before being called from on high to accompany Dante on that otherworldly journey. The sins punished from the second to the fifth circle are part of the very nature of humanity, and since they offend God the least, the relevant sinners have been assigned the upper part of Hell.

The second part of *Inferno* actually begins after the river Styx, "the river of hatred," which flows into the lagoon of the same name, Stygia, along which stretch the turreted walls of the city of Dis, redolent with a sinister glow of flame. This part includes two circles, the sixth and the seventh, having to do with bestiality: in addition to the Epicurean heretics, it accommodates those who were violent against their neighbors, i.e., tyrants, murderers, and marauders, who are submerged in the boiling blood of the river Phlegethon; those who were violent against themselves, the suicides, have been turned into brushwood, while those who were violent against things, the squanderers, run naked through the wilderness, pursued by ravenous black dogs that tear them to shreds once they catch them; and, finally, those who were violent against God and art, the usurers, are forced to sit in the red-hot sand under a rain of flames that fall from the sky like snowflakes in the mountains when there is no wind.

The third section, the deepest and darkest, is that of malice: this section welcomes sinners who have committed the most serious sin, that of fraud. The fraudulent have not lost their reason, but have instead exploited it to commit evil, a choice therefore wickedly conscious. Dante and Virgil descend the "steep slope" of this infernal cavern on the back of the monster Geryon, a winged serpent with a human face, while the roar of the Flegetonte waterfall echoes through the air.

The eighth circle, called Malebolge, is, in fact, a penal colony on a vast scale, with a very varied and articulated landscape consisting of ten moats divided by walls surmounted by small bridges, and is watched over by the devils of Malebranche, a group of diabolical beings led by Malacoda and adorned with colorful and bizarre names such as Alichino and Calcabrina, Cagnazzo and Draghignazzo, Graffiacane and Barbariccia.

The punishments to which these sinners are condemned amount to a veritable horror museum: in the first bedlam, pimps and seducers are forced to walk in opposite directions, whipped by horned devils, while the flatterers in the second bedlam are immersed in dung; in the third bedlam, there are simoniacs, stuck upside down in holes within the rock, the soles of their feet lapped by flames; the magicians and soothsayers of the fourth bedlam are forced to walk eternally backwards with their heads completely turned around; the barterers placed in the fifth bedlam are immersed in boiling pitch, and whoever tries to emerge is immediately seized by the hooks of the devils; the hypocrites in the sixth bedlam are condemned to walk while wearing heavy leaden cloaks gilded on the outside; in the seventh bedlam are the thieves, with their hands tied by snakes whose bite has the power to incinerate them, after which they are reconstituted exactly as they were before; the fraudulent counsellors in the eighth bedlam are wrapped in flames; The purveyors of discord and scandal are torn by the sword of a devil whenever they appear before them; those who have caused schisms are disfigured by foul diseases, such as scabies and leprosy; forgers of person, of coin, and of speech are also maimed by terrible diseases, such as rabies or dropsy.

Passing the well of the giants, one finally reaches the last circle, where the worst of all sinners are gathered: the traitors. Depending on the nature of their betrayal, these sinners are immersed differently in four areas of the frozen expanse of Cocytus: those who were traitors to their relatives are in Caina and are immersed up to their necks, with their faces turned downwards; traitors to the homeland, placed in Antenora, are in the same condition, but their faces are turned upwards; the traitors to guests are in Tolomea and are immersed in the ice in a supine position, with their faces turned upwards so that the tears freeze in their own eyes. And finally, in Giudecca, there are those who were betrayers of their benefactors, all completely immersed in ice in various positions: at the centre of the frozen lagoon is the horrid Lucifer, the origin of all evil in the world, who mangles with the jaws of his three faces the bodies of Judas, Brutus and Cassius, the traitor to Christ and the traitors to Caesar, symbols par excellence of the spiritual and imperial powers of the world. The icy cold that crosses the lagoon and freezes the waters comes from the mechanical movement of his six wings: like a hideous bat, Lucifer is there, stuck like a thorn in the centre of the Earth, in the heart of darkness of the world, the very emblem of evil, of monstrous envy and its futility.

Among the lost, one could not miss the Florentines; Hell is full of them. That infernal chasm is a sounding board for all of Florence's ills. Dante was not against the human pursuit of happiness, but he was absolutely against the idea that this could be achieved through wealth and power; looking around, he saw nothing but his fellow citizens lost behind the cursed florin, the easy wealth and the corruption that it brought with it:

'That race of newly rich, and rapid gains,
these seeds, Fiorenza, bring to flower in you
excess and pride. And you already weep for that.'

Inferno, XVI

All the sinners encountered in the pits of Hell are deprived of the vision of God and endure a perpetual torment for their sin that derives from the biblical principle of contrapasso. Dante does not describe sin, he shows us sinners directly, men or women moved by their cupidities, passions, or tragedies, and they are always portrayed in the round, whether Filippo Argenti or Vanni Fucci, Pope Boniface VIII or Francesca da Rimini, Pier della Vigna or his ancient teacher Brunetto Latini.

This underground land is the true theatre of crime and punishment, with the stage continually trodden by mythological creatures, half beast and half man, symbolising the world's dark moral dissolution: the Minotaur, the infamy of Crete itself, or the mad Harpies, the centaurs led by Chiron, Pluto, who utters incomprehensible phrases, or Geryon, the chimerical personification of fraud. All of these characters, with their accompanying mythological, legendary or historical events, become equally real and important in the surreal and infernal world created by Dante.

It is a polyphonic world made up of many distinct voices: they are abusive, sorrowful, petulant, blasphemous, sarcastic, kind and pathetic. The Malebolge is also the place where the comic and the ironic register, tinged with the grotesque, taking the upper hand with the saraband of the devils and the mockery of the sinner Ciampolo di Navarra, while the frozen swamp of the Cocito leaves only room for icy silence.

It's time to go further, as Virgil himself tells Dante: clinging to Lucifer's hair and doing an acrobatic somersault at one point to reverse direction (after all, the Earth is round!), they begin to climb towards the southern hemisphere:

We climbed, he going first and I behind,
until through some small aperture I saw
the lovely things the skies above us bear.
Now we came out, and once more saw the stars.

Inferno, XXXIV

The Hill of Purgatory

Purgatory is Dante's invention par excellence. Where patristics had shaped the visions of Hell and Paradise, there wasn't much material to rely on about Purgatory itself, even if logic dictated that there should necessarily be an intermediate place between irremediable sins and the eternal glory of heaven. It certainly could not be argued that all sins were of the same gravity, and then, what about those who repented at the end of their lives? Where to place them? After all, they were souls who had also seen the light, albeit in their last useful moments.

This serious geographical/theological shortcoming was first remedied by the Church, with the Council of Lyons in 1270, and then by Dante in literary terms, erecting the hill of Purgatory, the same mountain that Ulysses had glimpsed in his mad flight in the southern hemisphere before perishing miserably. In Purgatory, punishments are expiated through well-established accounting, sentence by sentence. Contrary to Hell and Paradise, places fixed in their wretched and luminous eternity, in Purgatory, everything is dynamic, everything is dominated by movement and the passing of time. If the sinners of Hell are immobile in their stinking pits, imprisoned by the chains of sin, the crags of Purgatory instead swarm with pilgrims in procession, like those heading for Canterbury in Geoffrey Chaucer's poem, or like the penitents along the Milky Way leading to Santiago de Compostela, all souls on a journey to rise ever higher towards heaven.

The purification of these souls in Purgatory is progressively perfected, step by step, with the same temporality of the earthly world, the end of time linked only to the advent of the Last Judgement. The rhythm is marked by day and night: here, Dante sleeps and dreams. In the realm of expiation, there is no longer the dark and distressing drama of Hell, but only enchanting dawns and serene sunsets. Human passions are diluted in the melancholic memory and in the humble recognition of past errors: everything is transparent, but as if muffled, like the view through the filigree of a linen curtain.

How did Dante and Virgil arrive in this otherworldly world? They landed on the island upon which the hill of Purgatory rises at dawn on March 27, 1300, passing through the "natural cave" attached to the hairy flanks of Lucifer. It doesn't seem possible to Dante that he has left behind that "dead aura" of Hell: he looks up as the starry sky begins to brighten and he notices that there is light on him now, there are shades of colour, while, before, there was only the livid gloom of sin. A hint of blue, the colour of oriental sapphires, tinges the celestial vault with serenity and heralds a world finally cleared of past anguish.

Also in the east shines crystalline Venus, the planet that invites humans to love, accompanied by four bright stars, invisible in the sky of the inhabited world, the stars of the shining Southern Cross: our pilgrim cannot help but interpret them as the four cardinal virtues. It is difficult to take one's eyes off that splendour of the firmament.... Before beginning his journey, Virgil collects dew from the grass and washes Dante's face with it, to remove every residue of infernal filth.

Dante in Crimea

Conversation on Dante, by the Russian poet Osip Mandel'štam is a classic of twentieth-century literary criticism. It was composed on the east coast of the Crimea, in the spring of 1933, when the poet spent an extended stay in a small town in the region. He had come to the Crimea with his wife while the almond trees were in bloom, with a "primitive's baggage," including a supply of bread. These were the dark years of the famine that had emptied all the granaries and markets from the Urals to the Don lands. The weeks he spent in Staryj Krym, behind the Black Sea, were the occasion for his total immersion in science—in particular chemistry, crystallography and natural history, the object of study of Buffon, Darwin and Lamarck—and in the poetry of the great Italian authors, Petrarch, Ariosto, Tasso and above all Dante. He recited the Comedy day and night, lightly rocking his body and cadencing the words. "He is a man full of rhythms, of thoughts, of words that travel," Lidija Ginzburg said of him.

Only a poet like Mandel'štam, with his unparalleled depth and passion, could restore the vivacity and vigor of Dante's verses in the pages of his Conversation on Dante. He found that there was a direct link that led him from Lamarck to Dante: in Lamarck's scientific scheme, the forms of life are organized hierarchically from the lowest and most rudimentary to the highest and most complex; the same hierarchy of values that he found in Dante. In his commentary on the Comedy he wrote that Dante had managed to "combine the incombinable". The brotherhood with Dante is total, he too an exile in his homeland and perpetually on the move. Mandel'štam was arrested and sentenced to five years' deportation for "counter-revolutionary activity" on 8 August 1938. His final destination is Siberia. The only letter he managed to write from the hell of the Stalinist gulags was addressed to his brother Aleksandr and to his beloved "little dove", his wife Nadinka. Then death, freed fhim rom the filth of the world and finally set him free to fly in the sky of poetry.

This second otherworldly realm is therefore situated at the centre of the southern hemisphere of the waters, at the antipodes of Jerusalem. At the foot of the mountain that dominates the island is the beach where the souls of those who have died in the grace of God are gathered: they arrive there directly from the banks of the Tiber on a very agile vessel guided by an angel.

The guardian of Purgatory was the pagan Cato, a strenuous defender of liberty, who committed suicide in protest against Caesar for the crime of having ruined ancient Republican Rome. The fact that he was both a pagan and a suicide did not preclude this austere Roman from exercising the role of guardian of that kingdom, because the taking his own life was an act of courage, not cowardice: as such, his figure had been glorified by all the great writers and poets of antiquity—Cicero, Lucan and Virgil himself—and had become the very symbol of freedom from tyranny and, by analytical continuation, from sin. And Purgatory is precisely the realm of freedom, freedom from sin, freedom from arbitrariness.

Geographically speaking, that kingdom of atonement for sins set on a mountain is a genuine fortified city, sometimes impenetrable, sometimes remote, divided into three parts. The Antipurgatory occupies the lower part of the mountain, the actual Purgatory the central part, and the Earthly Paradise the upper part. The atmosphere envelops the lower part up to the door of Purgatory: from there on, up to the top, there are no more atmospheric disturbances, no more clouds or rain, no climatic cataclysm to disturb the serious serenity of those places.

As if following Saint Thomas' structure and classification of the vices of misguided love to the letter, Purgatory is divided into seven frames, in which the seven deadly sins are expiated: pride, envy, wrath, sloth, avarice, gluttony and lust. Everything is antithetical and specular with respect to Hell: the funnel-shaped depth of Hell corresponds to the conical mountain of Purgatory; while the former sinks into the bowels of the Earth, the latter raises its peak in the opposite hemisphere; and finally, if, in Hell, as Dante progressively descended, he met the worst sinners, here, proceeding upwards, he passes from the most serious sin to the lightest. Moreover, as we ascend, the terraces widen, having, as an inner wall, the rocky part of the mountain, but in front of it, the boundless view of the sea. Each frame has its own guardian angel, and sinners have examples before their eyes of the vice punished and the virtue opposed.

In the lower part of the mountain, in Antipurgatory, are the lazy, the excommunicated, those who died from violent death, and the negligent princes. For all these sinners, ascending to the kingdom of heaven is only a matter of time. The lazy, for example, must wait in Antipurgatory for a time equal to their lives, and only then can they begin to climb the various frames of Purgatory: if, in life, they were haughty, they will climb to the frame of the proud; if lustful, they will head instead to the frame of those who were dominated by sensuality. For the excommunicated, the accounting of the punishment is manifested in a more articulated manner: thirty years must elapse for each year of excommunication. All these wretches willingly subject themselves to sufferings—as painful as those endured by the truly damned in Hell—because they have the hope that the stain of their sin will be removed once the time of punishment has passed. Those who have appropriately turned to God at the moment of their death by intoning the psalm Miserere Domine, such as Manfred, Frederick II's illegitimate son, or the Ghibelline leader Bonconte da Montefeltro, know that the glory of Paradise awaits them at the end. After passing through the Antipurgatory, where he has met an old Florentine friend, the musician Casella, with whom he sweetly sings "the amorous song that used to quench all my sorrows," Dante climbs laboriously with Virgil, who tells him that the ascent to the mountain, whose summit cannot be seen, will be more laborious at first and become easier later.

Among the many penitents he meets on the way up, here comes Sordello, from the same Mantuan land as Virgil, addressing spontaneous and fraternal words to the Latin poet, moved by a great feeling for his country. Seeing all of this, Dante bursts into a fierce invective against the Italy of the time, in the hands of tyrants and subject to the whims of the pope:

You! Vile Italia! Sorrow's resting place!
You hulk that no one steers through raging storms!
No sovereign lady, you're a cat house whore!

Purgatorio, VI

Sunset is approaching, bringing with it an atmosphere of poignant melancholy.

After having spent the night in a fabulous "little valley" with a thousand coloured flowers and populated by illustrious personages who did not know how to make good use of their power in life, at dawn, Dante dreams of being abducted by a great eagle that lifts him up to the sphere of fire, and the impression is so strong that he wakes up and finds himself in front of the door to Purgatory, watched by an angel. Dante humbly asks him to open that door for him: the angel then carves seven P's into the Poet's forehead with his sword, exhorting him not to look back and to wash away these marks as he ascends the frames of Purgatory. Then, using two keys, one silver and one gold, he lets him in.

Those seven marks represent the seven deadly sins that compel Dante to continue in his ascent by engaging in a struggle inside and outside of his body to purify his soul. Each of those marks disappears as a frame passes, and Dante arrives at the top of the mountain a new man, as pure as Adam before committing the original sin. Unlike in Hell, where the damned all have a great desire to shed their anonymity and have their say publically, even if their story was cruel, in Purgatory, the liberation from the corporeal envelops all of the characters like a sad and melancholy fog. There are no longer clashes or violent dialectical exchanges like those with Filippo Argenti or Farinata degli Uberti, but only meetings and conversations, as in a continuous monologue of a soul in search of itself. So, the lexicon changes, the words become attentive to light and sweetness, the tone is medium and elegiac. "The noun," notes Osip Mandel'štam, "becomes the end, the purpose, rather than the subject of the sentence."

The passage of Dante and Virgil through each step of Purgatory follows an inexorable logic that is repeated to the letter: upon arriving at a new frame, Dante (the narrator), using various literary devices, presents virtuous examples antithetical to the sin that needs to be purged; then, before leaving that frame, he mentions various characters who have been punished precisely for that vice, ultimately proclaiming their beatitude. For example, in the frame of the envious, which is the theme of Canto XIII of *Purgatory*, the two poets hear invisible spirits flying around shouting examples of charity: the first spirit repeats aloud "Vinum non habent," or the words spoken by the Virgin Mary at the wedding in Cana; the second shouts "I'm Orestes," the words spoken by Pilades as he died in his friend's place; the third shouts "Love those by whom you suffer harm," the same words spoken by Christ to the Apostles. The sinner of the Canto is Sapìa, a Sienese gentlewoman so envious of her fellow citizens that she wished that, in the battle of Colle di Val d'Elsa, which took place in 1269 between the Sienese and the Florentines, the Sienese would lose.

Canto XVI of *Purgatory*, dedicated to the wrathful, has a very special place, because not only is it the central Canto of that Canto, but also that of the entire Comedy. And what is it about? It is about the problem of free will, which is the basis of the entire work, since, for Dante, denying the free will of humanity alters the entire system of punishments and gratifications that result from human actions.

In Purgatory, there is also the continuous presence of fire: the eagle that kidnaps Dante in the dream raises him to the sphere of fire, while, at the end of the ascent, there is a real wall of fire that Dante must cross, as it divides him from the vision of Beatrice:

And, once within, I could have flung myself –
the heat that fire produced was measureless –
for coolness, in a vat of boiling glass.

Purgatorio, XXVII

Once they have climbed all the steps of the mountain, the two poets set out up a long staircase that leads to the top of Purgatory, where the Earthly Paradise is found. However, night falls and, unable to continue, they stop to sleep on a step between two high rock walls. Dante sees the stars shining in the sky, then falls asleep and dreams. A beautiful young woman appears to him, intent on picking flowers and making garlands of them. When he wakes up, Virgil tells him that a day of great joy has dawned for him, because they are close to the Earthly Paradise, where he will leave Dante, entrusting him to Beatrice's care.

And so, on the morning of March 30 of the year 1300, he arrives in the Earthly Paradise, an immense expanse of grass and flowers where a delicate perfume hovers everywhere and the air is full of the chirping of birds. Going into the woods, he ultimately arrives on the banks of the river Letè, with its clear and pure waters, and on the opposite bank appears a young woman, Matelda, intent on collecting flowers: she tells him that the river Letè, to those who immerse themselves, erases the memory of sin, while the other river, Eunoè, revives the memory of the good done on earth.

The ending of *Purgatory* is impressive: Dante thinks he sees a host of golden trees among the trees of the Earthly Paradise, but, getting closer, he recognizes that they are actually seven gigantic golden candelabra, each with a vivid flame at the top that, as they move, leaves its own rainbow trail. Behind the candlesticks is a procession of spirits, all dressed in white: there are twenty-four old men, their heads encircled with lilies, followed by four large animals, each with six wings and adorned with green fronds. They are a lion, a bull, a strange beast with a human face, and an eagle. They all proceed alongside a triumphal chariot pulled by a majestic griffin, its body half that of a golden eagle and half that of a red and white lion. To the right of the chariot dance three maidens dressed in different colours, to the left, a number of other maidens, their leader being a woman with three eyes. Finally, a further group of old men, seven in all, closes the procession.

And behold, from a cloud of flowers scattered by angelic hands, a woman appears, dressed in red, wrapped in a green cloak, her head covered by a white veil. It is Beatrice. Dante does not recognize her, but he senses that it is her, so he turns to Virgil, to tell him how pained his heart is: but Virgil is no longer there! His eyes fill with tears while a sweet voice whispers:

'Dante, that Virgil is no longer here,
do not yet weep, do not yet weep for that.
A different sword cut, first, must make you weep.'

Purgatorio, XXX

And then, that woman introduces herself to him as Beatrice and asks him how he could think himself worthy to climb up to that mountain, reproaching him because, after her death, Dante had forgotten her and had abandoned himself to a life of sin, descending lower and lower into the scales of vice, until he was lost in a dark forest. It was then that Beatrice had prayed to God to grant him salvation.

Shaken by so many revelations, Dante is struck by the power of ancient love and faints; when he comes to, he finds himself in Matelda's arms, immersed in the waters of the Letè. Then, he hears Beatrice's voice ordering Matelda to immerse him in the waters of the Eunoè as well, so that he could remember only the good that he had done:

I came back from that holiest of waves
remade, refreshed as any new tree is,
renewed, refreshed with foliage anew,
pure and prepared to rise towards the stars.

Purgatorio, XXXIII

The Light of Paradise

With Dante's flight towards the heavens of Paradise, accompanied by Beatrice, the third Canticle begins: it is 12 noon on March 30, 1300, the time of the most favourable conjunction of the sun, the "lamp of the world," with the best brilliant constellation in the firmament, that of Aries.

As we know, everything in the Comedy has a symbolic value. And if Dante had lost his way in the dark forest at dusk (night, as a symbol of the soul abandoned by God), if he had then arrived on the beach of Purgatory at the first light of day (dawn, as a sign of the redemption of one's sins), here, he is now ascending into Paradise when the sun is at the height of its triumph, and the soul is therefore happier to be reunited with God:

Glory, from Him who moves all things that are,
penetrates the universe and then shines back,
reflected more in one part, less elsewhere.

Paradiso, I

The last part of Dante's journey through Paradise lasts about twenty-four hours, almost all in the company of Beatrice and her smile: only in the last stretch will Beatrice be replaced by Saint Bernard of Clairvaux, the most ardent of mystics.

If *Inferno* is the Canticle dominated by characters, *Purgatory* and, above all, *Paradise* are instead the Canticles dominated by scientific, philosophical and theological discourses, learned explanations that Dante receives from his beloved and from the other chosen ones he meets during his passage. *Paradise*, moreover, is the Canticle of light, a light that radiates everywhere, that shines both in Beatrice's eyes and in the rotating sphere of the heavens and that becomes all the brighter and more dazzling the higher one ascends towards the vision of God.

This time, it falls to the Poet to perform an arduous, almost impossible and paradoxical task: to describe in words what cannot be stated in words, since Paradise is ineffable, excludes comprehension and representation, and can be grasped only through faith. Primo Levi wrote that our language is too limited to describe the motions of the planets and the invisible motions of atoms, just as there are no words to describe the abyss of Auschwitz. Dante has, in a sense, the opposite problem, that of having to invent and bend words to speak of a reality that transcends it.

It is within this antinomy that all of the linguistic difficulty of the third Canticle resides, given that, if words were able to represent the nature of God, it would cease to be of a transcendent nature, mysterious and beyond human. Approaching Paradise and its vision, therefore, means going beyond human experience, detaching oneself from the senses and external reality, abolishing space and time, probing only the essential silences of the spirit, touching with one's own hand the poignant inability to describe a truth incommensurable with our senses:

> *To give (even in Latin phrase) a meaning*
> *to 'transhuman' can't be done. For those whom grace*
> *will grant experience, let my case serve*
>
> Paradiso, I

The structure of Paradise is, like that of Hell and Purgatory, very articulated, but, unlike the earthly narrowness of the other two realms, here, all the richness of the celestial spheres unfolds, according to the order dictated by the Ptolemaic planetary system, with immaculate and limitless landscapes: the Earth stands still in the Universe, and around it revolve nine heavens (Moon, Mercury, Venus, Sun, Mars, Jupiter, Saturn, Starry Heaven, First Mobile). The tenth heaven, the Empyrean, envelops them all and is motionless; it is there that God resides, surrounded by nine festive and resplendent angelic choirs together with all the blessed souls arranged in the form of a candid rose.

All of Paradise is made up of light and music, of order and harmony, and it shares in the immanent presence of God throughout the cosmos. The souls of the blessed carry their earthly history with them, however transfigured: they are more or less close to God according to their merits, but they go with Dante from heaven to heaven (from that of the Moon to that of Saturn) to show their different degrees of bliss and to allow him, with their splendour, to refine his visual abilities so that he can, in the end, sustain the vision of God, source of light blazing in the cosmos and central point of the heavenly Rose.

Each sky is characterised by a particular moral virtue: the Moon's sky is characterised by strength, Mercury's by justice, Venus's by temperance, the Sun's by prudence, Mars' by faith, Jupiter's by hope and, finally, Saturn's by charity.

We now come to the main stages of this last journey, starting from when, after having left the Earthly Paradise and crossed the sphere of fire, Dante and Beatrice rise with very rapid movement towards the first sky of the Moon, where the spirits of those who were forced by the violence of others to break their religious vows are to be found. Here, Dante meets Piccarda Donati, who was abducted from the monastery where she lived and forced into marriage for political reasons by her notorious brother Corso.

Passing into the sky of Mercury, where those who used their wits to do good hover, Dante meets the great emperor Justinian, who celebrates the highlights of the Roman Empire, from Aeneas to Charlemagne. Amidst disquisitions on the death of Christ, the redemption of man and the incorruptibility of things created by God, Dante and Beatrice arrive at the sphere of Venus, where they meet Charles Martel and other blessed ones. The sky of the Sun is that which shines with the wisdom of Saint Thomas and Bonaventure of Bagnoregio, with the praise of the faith of Saint Dominic and Saint Francis.

In the fifth heaven, there is a large scene featuring all of the blessed who died fighting for the faith arranged in the form of a luminous cross: from the right arm of this cross is revealed Cacciaguida, Dante's great-great grandfather who fell in the second crusade. The Poet then delivers a monologue focused on the noble virtues of ancient Florence, now disappeared, predicting his own painful exile as well.

Leaving the sixth heaven, that of Mars, they arrive in the seventh heaven of Saturn, where they witness another admirable choreography, with the spirits of the chosen ones arranged according to a very long ladder that climbs towards the Empyrean. Saint Pier Damiano speaks of the mystery of predestination, while Saint Benedict tells of the monastic order that he founded and laments its decline.

As Dante ascends into the various heavens of Paradise, physical space disappears around him to make way for the whirling sparkle of the blessed spirits, who shine like flaming sparks, like lanterns of light in magical movement, like birds when they fly happily in a flock. Having entered the eighth heaven of the fixed stars, Dante lowers his eyes and sees, very distant and petty in its smallness, the land from which he departed. Then, he lifts his eyes to Beatrice and higher, where he witnesses the triumph of Christ and the procession of the blessed.

He feels that he has arrived at the roots of wisdom, and this imbues him with a feeling of fulfilment. The conquest of happiness has come at a high price: he left full of tears from the gate of Hell, he climbed the crags of Purgatory with difficulty, and now he is there to enjoy the light of God and the unspeakable beauty of his beloved. And when Beatrice entrusts him to Saint Bernard, Dante understands that he will no longer be able to proceed by the force of reason, but only through the raptures of faith.

Invoked by a touching prayer by St Bernard, the Virgin Mary intercedes for Dante, granting him the sublime grace of the ultimate vision of God. Everything is consumed in an instant, one that no memory can capture: Dante sees the mystery of the Universe, the supreme truth of the world. He has finally reached the end of his journey; now, all that remains is for him to return to earth and recount it.

The Torments of Botticelli

There is no one in Europe who has had a greater artistic influence than Dante. And we are not just referring to poets or writers; we are also talking about painters and the major protagonists of figurative art from the thirteenth century onwards. Ever since the circulation of its first copies, the *Commedia* had fascinated illustrators and painters, as testified to by the great number of illuminated manuscripts and frescoes found in cloisters and churches.

By the end of the fifteenth century, several Italian artists had lent their hand to the illustration of certain parts of the poem, and we want to mention here, in particular, the singular story that links Dante Alighieri to Sandro Botticelli, the painter of the Primavera and the Birth of Venus, paintings that are the very symbol of the Renaissance.

Alighieri and Botticelli, the poet and the painter. The two could not be more dissimilar. Their lives were separated by about two hundred years, an abysmal lapse of time, in which the Florence of the Guelphs and Ghibellines had been replaced by the Florence of Lorenzo the Magnificent. If the former lived in a city politically torn apart by civil war and culturally dominated by the philosophy of Saint Thomas (with some flashes of Averroism), the latter had had the good fortune to grow up in a Florence governed in an enlightened manner by the Medici, when the city had become the beating heart of the Renaissance: the Medici court, made up of bankers, financiers, savants, painters and poets, architects and urban planners, neo-Platonists and Epicureans, was a hotbed of new ideas, destined to revolutionize art and change the view of the world.

Dante was the very image of the medieval sage, versatile in all of the arts and in the fields of theology, philosophy, and science; Botticelli, on the other hand, was a fragmentary artist and, according to Giorgio Vasari, also poorly educated. If, for the former, the exemplary woman is the chosen Beatrice, for the latter, it is the pagan Venus, with all the corollaries associated with this magnificent female ideal.

Yet, life had a surprise in store for Botticelli, a surprising turning point.

It all began with the death of Lorenzo il Magnifico in 1492, and with the appearance on the Florentine political-religious scene of the preacher from Ferrara, Girolamo Savonarola. Savonarola was a Dominican friar obsessed with pagan culture and the lascivious customs of the time, which he attacked vehemently in his sermons in the piazza, heedless of accusing rulers and prelates in the impetuosity of his sermons.

The Church was then led by the Spaniard Rodrigo Borgia, a pope with whom the Holy See of Rome had reached the bottom of dissolution: "There is nothing good in the Church, from head to toe," cried Savonarola, "this is not the holiness of God. In lust you have become an impudent harlot, you are worse than a beast, you are an abominable monster." He immediately attracted the favor of the simple, of the poor commoners, but also the sympathy of those who opposed the Medici family.

He conquered the Florentines with his passionate sermons, his rhetoric full of quotations from the Apocalypse and the Gospels, the thundering voice with which he asked Christians to do penance for the imminent arrival of the Last Judgement. While his most fanatical followers organized themselves in the penitential sect of the "whiners" (so called because of the tears shed during Savonarola's sermons), he himself organized the "pyres of the vanities," where works of art, books and musical instruments were burned, guilty of having corrupted the souls of the faithful.

Botticelli was deeply shaken by Savonarola's sermons and his impetuous words. He began more and more often to brood over them and, in so doing, triggered in himself a great sense of guilt for all his artistic production. An intense mystical-religious crisis began, which led him to the darkest of melancholy, to the saddest of thoughts. He became austere, restless and solitary. But he was still one of the most important painters in Florence, and when Lorenzo di Pierfrancesco de' Medici, known as Lorenzo il Popolano, second cousin of the much more famous Lorenzo de' Medici, wished to have a sumptuously illustrated manuscript of the *Commedia* in 1480, he could only entrust it to Botticelli.

The encounter with Dante and his divine poem changed Botticelli's life forever: he devoted ten very intense years to preparing the illustrations for the *Commedia*, a work for which he burned and trembled, and which caused him "endless turmoil," if we are to believe Vasari's words. He began an enormous graphic work that remained unfinished: about a hundred drawings on parchment representing a reconstruction of all the Canti, plus the infernal chasm, the only truly complete panel.

The drawings, for the most part, were created with a lead stylus, after which he would add ochre or black ink to give them depth. He proceeded slowly, as if obsessed by the work, and followed the text to the letter, trying to reproduce the incisiveness of the Poet's verses with his yellow gold strokes. With Botticelli, we are, for the first time, faced with something totally new compared to the illuminated manuscripts that had accompanied the Canti della *Commedia* in the past: his work is, in fact, a true modern-style graphic novel.

With his particular artistic sensitivity, Botticelli was able to penetrate deeply into the souls of all the characters in Dante's poem, illustrated in minute detail in the instant of the torment of their suffering or in the bliss of their joys. In these panels of the highest quality by Botticelli, acting as a faithful mirror of all Dante's mastery and his parable as a man and an artist, the Poet ascends from Purgatory to Paradise as if almost flying, the crude scenes of Hell disappear and the figures become increasingly abstract and impalpable.

It is a unique artistic corpus, but, after the artist's death, a real detective story began, linked to the disappearance of those precious plates that only reappeared hundreds of years later in two different countries and under unusual circumstances.

It was 1882 when the Times of London announced the sale at auction in Edinburgh of the most important ancient manuscript ever to appear on the market. In order to cover all of the debts that he had accumulated from horse racing and gambling, William, Duke of Hamilton, had been forced to put up for sale, together with the rest of the family treasures, a series of plates that his grandfather had bought in Paris at the beginning of the nineteenth century from a certain Gian Claudio Molini, a renowned collector and printer: they were the precious Botticelli parchments. How Molini got hold of them remains a mystery to this day: some hypothesize that he bought them in Florence and then took them to France, while others suspect that he found them in France, where they may have arrived as a gift from their legitimate owner, Lorenzo il Popolano, to King Charles VIII.

At the news of the auction, Friedrich Lippmann—expert curator of the State Museums of Berlin—came forward with a very generous offer. Sensing the extraordinary bargain, he had convinced the German government to take charge of the purchase. At this point, motivated by greed, the Duke of Hamilton had no qualms about proposing the following deal to Lippmann: acquire the Botticelli panels without participating in the auction, provided he was willing to purchase the entire collection. Pressed by Lippmann, the German government gave in and accepted the Duke of Hamilton's conditions, purchasing all of the manuscripts in his possession for an astronomical sum.

However, when the zinc boxes arrived in Berlin from Scotland, it was discovered that there were only eighty-five plates inside; seventeen were missing. The origin of the problem was that the set of plates had already been divided up in the past, as evidenced by the purchase of another seven Botticelli parchments in 1632 by the Queen of Sweden from two Parisian collectors. The Swedish sovereign, following a serious religious crisis, converted to Catholicism, abdicated and moved to Rome, in the wake of which, in 1690, she gave the plates to the Vatican. And these seven panels depicting the Comedy executed by Botticelli are still there today, in the archives of the Vatican Museums.

Ten more plates are therefore missing, and no one knows where they have ended up. Botticelli's drawings have therefore partly followed the fate of the text of the poem that had so fascinated the painter, perhaps scattered among a number of mysterious private collections or perhaps lost forever.

A Non-existent Book

It was because of two different naval battles, that of Meloria in 1284 and Curzola in 1298, that Rustichello da Pisa and Marco Polo found themselves both prisoners of the Genoese and sharing a prison cell.

The naval battle fought in August 1284 along the Tyrrhenian coasts of the Meloria Tower put a definitive end to the Maritime Republic of Pisa's dreams of glory: the bulk of the Pisan fleet was destroyed by the Genoese and sank in the waters of the Tyrrhenian Sea, in a ruin of floating ships, bodies and blood. Thousands of Pisan sailors were captured, which left the republic's male population greatly diminished, leading to progressive decadence. It was on account of this large number of prisoners that the saying was born, "If you want to see Pisa, go to Genoa." Among that unhappy bunch was the Pisan storyteller Rustichello.

In the summer of 1298, while the city of Florence was inflamed by the increasingly heated political struggle between the White Guelphs and the Black Guelphs, the Adriatic Sea was preparing to be the scene of war between the two oldest and most powerful maritime republics, Venice and Genoa, who had long been stuck in a vortex of provocation and mutual retaliation. Genoa had entrusted Admiral Lamba Doria with command of an imposing naval fleet, composed of 85 galleys with very skilled crossbowmen on board. They were tasked with wreaking havoc right under the nose of the Serenissima and raiding the Venetian colonies in Dalmatia, both on the mainland and on the islands.

Provoked in such a mocking way at home, Venice hastily prepared a fleet of 98 warships under the command of Andrea Dandolo, also an expert sea dog. The clash between the two naval armies took place on September 7, 1298, and was particularly violent and bloody. Playing cunningly, the Genoese commander had detached 15 ships of his fleet the day before, stationing them behind the nearby island of Lagosta, with the order to pounce on the Venetian ships once the battle had been engaged, as they would enjoy the advantage of the favourable wind. And so it happened: the unexpected arrival of the Genoese galleys, with the rain of darts and incendiary arrows that came from their decks, threw the Venetian fleet into disarray, leading to them suffering one of the most catastrophic defeats in their history. As many as 80 Venetian galleys were captured by the Genoese: the Venetians reported 7000 dead and 8000 prisoners; among the dead was the commander Andrea Dandolo, who had preferred to kill himself rather than be held prisoner. Among the prisoners was a forty-four-year-old commander of one of the galleys, Marco Polo.

Rustichello da Pisa and Marco Polo found themselves sharing a large room in the oldest part of Palazzo San Giorgio, near the docks of Genoa's old port: from the window of their cell, they could hear the voices of the sailors, the crash of the boats as they docked, the thud of the sacks unloaded from the galleys. According to Giovanni Battista Ramusio—the great Venetian publisher who, in 1550, published the monumental geographic treatise Delle navigationi et viaggi, which compiled more than fifty memoirs of voyages and explorations—during the months of his detention in the Genoese prisons, Marco Polo was given the white glove treatment:

> *The fame of Mark's great virtues, of the virile way in which he had fought, of his portentous travels, had come before the city [Genoa] and although his freedom was not restored, he was kept not as a prisoner, but as a dear guest. To see and hear so rare a man, the whole city came to his prison. Everyone was eager to know the wonderful things he had seen, and forced to repeat them several times in one day, which gave him tedium and discomfort, he was advised by a noble Genovese friend of his, to put in writing the report of his journey, and had the opportunity to bring his memoirs and writings from Venice in that same year, dictated it in French to a Pisan said Rustichello, companion of his injury.*

> **Ramusio, the Editor's Touch**
>
> In the first half of the 16th century, in the rich panorama of Venetian Humanism, a singular intellectual figure emerged: Giovanni Battista Ramusio, secretary of the Senate of the Republic of Venice, geographer but above all publisher. A subtle diplomat, very attentive to the changes brought about in Europe by the widening of geographical horizons, Ramusio skillfully inserted himself in the world of cartographers, collecting testimonies and notes of explorers, navigators, merchants and adventurers. Thus were born three large volumes of a work that forever immortalizes his name: Delle navigationi et viaggi, published in various editions between 1550 and 1606, a gold mine on every corner of the new and ancient world, the result of the experience and daring of great travelers. The collection includes over fifty travel accounts and geographical treatises, including the discoveries of Amerigo Vespucci, the voyage of Magellan, the chronicles of Pietro Martire d'Anghiera on the discovery of America by Christopher Columbus and a rare and extensive edition of Marco Polo's travels, reconstructed by Ramusio on the basis of a manuscript he found in Venice. The title Il Milione is due to Ramusio, for reasons that remain uncertain: did he perhaps intend to refer to the wealth of Kublai Khan? Or to the nickname of the Polo family? Or to the countless fantastic things described in the text? Doubts that have never been clarified. Ramusio was the true creator of Marco Polo's imperishable fame, starting with the picturesque description he gives of his return to Venice after an absence of more than twenty years. Although two hundred years have passed since that episode Ramusio has the ability to describe it with the consummate art of a chronicler present at the scene:
>
>> He wrote that the three Polos had disembarked from a galley, staggering about like old sailors, dressed in very heavy, shabby tunics, bound at the sides by thick cords, and with dirty shoes on their feet. They had a certain indescribable Tartar trait in their appearance and accent. When they arrived home, they gave a great feast in the evening to celebrate their return, to which they appeared dressed in splendid Venetian red togas. In front of their guests they took the clothes in which they had travelled and cut them with a blade, causing rubies, emeralds, diamonds, lapis lazuli to fall from the folds of the seams, which began to roll on the floor, to the great surprise of all present.

Rustichello was pleased to share a cell with such a person while thousands of his Pisan brothers continued to die in the dungeons of Genoa, subjected to very harsh imprisonment. He found that he could kill the monotonous flow of time by doing the only thing he knew how to do: write. The two cellmates subsequently found a way to create a manuscript that remains one of the most extraordinary and fascinating travel and adventure books ever written, *Il Milione*, also known as *The Book of the Wonders of the World*.

Thus, by one of the many subtle paradoxes of history, the narrow space of a Genoese prison cell in 1298 provided the opportunity to reveal to the entire West the geographical vastness of Asia, the colours of the endless Chinese deserts, the perfumes of Samarcanda, the customs and traditions of the Tartars, the buzz of the Arab ports, the wonders of the Far East, the echo of legends and magic, and the riches of medieval China. Marco Polo lists between the lines the countless goods desired by the Europeans of the time, from the silver of Armenia to the finely wrought silks of Persia, from the pearls of Baghdad to the golden carpets of Tabriz. rubies and sesame oil from Central Asia, ebony chess from Siam, cotton and silk from Catai, asbestos from Taklamakan, ginger and cinnamon from Bengal, indigo sandalwood and ambergris from Zanzibar, pepper and cloves from Java.

The original title of the book, Le Divisament dou Monde, or The Description of the World, well expressed its ambition to be a cosmographic synthesis of time, a complete account of the vast Asian continents and the lands from Persia to China, travelled far and wide by Marco Polo over twenty-four years of travel. In one fell swoop, it encompassed a number of medieval literary forms, serving as an encyclopedia, a scientific and geographical treatise, a short story and a mercantile account: the book was all of this and more, because it was a mélange of these genres.

Marco Polo and Rustichello: never in literature has there been such a happy encounter, such a perfect osmosis of two such different and contrasting worlds. The first was a Venetian merchant and a great traveller, at ease in handling diplomatic and mercantile dispatches, administrative reports and accounting notes, the latter a medieval novelist and courtier, accustomed to writing stories of chivalry and fantasy, such as the legends of the Breton cycle of King Arthur, that is, events of pure entertainment for the receptions of the various European courts.

Within the confines of that cell, and relying on the memory of things seen and heard, the great Venetian traveller returned with Rustichello to the boundless and polychrome spaces of Asia, to its immense rivers, to the tingling cities of Catai, to the unreal bejeweled palaces of Baghdad, to the most varied exotic animals, to the customs of those remote peoples and to the heart of the great Mongol Empire, the court of the much loved and revered Kublai Khan.

Rustichello was happy to give voice to the admirable adventures of "Messer Polo" and, thanks to his pen, the travel notes of the Venetian was transformed from an austere chorography of the new continents—almost intimidating for its list of countless cities and the amount of information and details of places, villages and markets—into a myriad of stories and fascinating tales. The result was, more than an articulate geographical treatise or a colourful travel book, something much more important: an expression of the yearning for discovery and the most extreme curiosity, the enthusiasm of travelling, the desire to get to know the new and immense spaces of the East, places situated at the end of the then-known world, countries that, for the medieval reader, remained mysterious and legendary. Marco Polo's enchanting gift to medieval Europe was the breath of great geographical horizons. As the incipit of *Il Milione* states:

> *Gentlemen, emperors and kings, dukes and marquises, earls, knights and bourgeois, all of you who want to know the different human singularities in the different regions of the world, welcome this book: read it or have it read to you.*

In the tale of this fortunate literary couple, fable and reality found themselves united in an articulated and composite text, outside of any other literary tradition, in which innumerable styles—from the novella to the exemplum, from the hagiographic tale to the historical account—merged into a single incomparable narrative. The tone, it must be said, is generally bare and marked by understatement, but what makes for compelling reading is the stupendous sequence of fantastical cities, full of rich markets and with magical names such as Aleppo, Tabriz, Kirman, Baghdad, Samarkand, Semenat, Chinsai, Cabaluc… Cities linked to memory and desire, mixed with the smell of musk or incense, the fragrance of magnolias and lavender:

> *Cities, too, think they are the work of the mind or of chance, but neither is sufficient to hold up their walls. You do not enjoy the seven or seventy-seven wonders of a city, but the answer it gives to one of your questions.*

The description of the cities is followed by the story of the evening camps, the smell of the elephants after the rain or of the damp sandalwood in the braziers, the hunting scenes and battles, the magical and fantastic events in which Persian haruspices, Uighur magicians or Nestorian priests stopped time or predicted the future. Amidst the description of villages and ports, the where and when often gives way to notes and comments on local products, digressions on historical and anecdotal events, descriptions and notes that still amaze for their depth and ethnographic acumen.

Marco, ambassador of the Great Khan, was interested in seeing and recounting "the many wonders of the world." Passionate curiosity, a sense of the unknown, a thirst for the new and the unusual aspects of reality: these were—as with Dante's Ulysses—the distinctive traits of our Venetian traveller and the guiding themes of his admirable work.

If the literary union that took place in a Genoese cell has all the traits of a singular event, equally curious is the history of the manuscripts and the various printed editions. Marco Polo and Rustichello's originals were lost almost immediately, as was the case, years later, with Dante's *Commedia*. *Il Milione* is therefore a text that simply does not exist.

The Merchant and the Poet

Nevertheless, it has survived the passage of years, indeed, centuries, being handed down and transformed into countless editions and translations, copies of copies, each with its own voice and its own character. With the multiplication of the copies, there was the inevitable rapid appearance of variants, even significant ones, with even the title of the book itself being subject to transformations: to sum up, at least one hundred and forty-three versions of the book have been counted, in almost all European languages, from Catalan to French, from Tuscan to Franco-Venetian, the oldest of which seems to date back to 1351 and is probably the one used by Ramusio for his work.

For *Il Milione*, paraphrasing a suggestive text by Luciano Canfora, it's really more appropriate to talk about the "copyist as author." And is there any other book for which the following considerations by Canfora are better suited?

> *In the universe of texts, there are "originals" that have never existed. Or, to put it better, texts at the origin of which there has never been an "original" and which have nonetheless lived and prospered. [...] The texts live separately from the authors who set them in motion. The only "omnipotent" author is therefore the copyist, the one who writes the text. [...] The changes made by him, those intended and those not intended, enter into the body of the writing and remain there. And as they make the writing they can also make history.*

And while this polycentrism gave many a headache to those who undertook the difficult philological task of reconstructing the original text (perhaps discovering its absence in the end), it also marked the rapid diffusion of the restless and magical world of thirteenth-century Asia into European culture. The procession of copyists and translators that littered the late Middle Ages was characterized by the fact that they all felt compelled to alter and embellish the text as they pleased, following their instincts, the interests of their masters, or those of the environment in which they worked.

The final result of this mixture of instincts is the polyphonic babel of the book itself: it must be acknowledged once and for all that the text with which we are now familiar has not a single author, but dozens of co-authors. It is the collective result of a large number of chroniclers, scribes and editors who have overlapped over the years, in simple terms, a Wu Ming Foundation ante-litteram.

Moreover, Le Divisament dou Monde was originally written by Rustichello in a French-Mediterranean language—the language spoken in the fourteenth century throughout the Mediterranean basin and bazaars, by European and Arab merchants, slaves from Malta and Maghreb corsairs—later translated into French, and only eventually transcribed into Italian!

The most avid bibliophiles, aware of the nuances and differences among the various copies, obviously owned more than one copy. Charles V, for example, had five in his library, while his brother, the Duke of Berry, owned three. And one of these is among the most prestigious examples of *Il Milione*, a princely gift given to the Duke of Berry in 1413 by his nephew, the Duke of Burgundy, known as John without Fear.

The Art of Copyists

The most prestigious copy of the Book of the Wonders of the World was commissioned by the Duke of Burgundy and was the opening book of a collection of seven other treatises on the Orient: the subject was of particular interest to the duke, because he had participated in the crusade against the Ottomans, had been taken prisoner in 1396 and was finally freed thanks to Tamerlane's victory over Bayezid in 1402. The collection included, in addition to Marco Polo's text, the writings of Odorico di Pordenone, John Mandeville and others. On the death of the Duke of Burgundy, the book passed to his son, John Fearless, a cruel and unscrupulous politician, who subsequently gave it to his uncle, the Duke of Berry. The volume then came into the possession of the Armagnac family, but, in consequence of the reverses which this family met with, and the dissolution of its library, it was long considered lost.

It was found in 1622 by Nicolas Rigault in the library of the King of France while he was making an inventory. Currently preserved in the Bibliothèque Nationale de France in Paris, the Duke of Berry's copy, the work of two copyists and a group of artists from the workshop of the Master of Boucicaut, consists of 84 splendid miniatures, with an almost dreamlike charm that has never ceased to seduce those who have had the privilege of contemplating them.

Thus The Million ended up having an enormous influence on the development of geographical maps and subsequent discoveries. Until the last days of his life, Christopher Columbus was under the illusion that he had reached not the Caribbean beaches, but rather the marvellous islands of the Asian seas and the coasts of Japan, that mysterious Cipangu of which Marco Polo spoke. This is the undisputed power and fascination of the book where "the wonders of the world" were told and, at the same time, our excuse to talk about them.

Diplomatic Missions

There had already been direct contacts between the great cultures and civilizations of East and West before Marco Polo's voyage. We are not talking about the commercial exchanges that took place along the Silk Road and then found their way to the ports of the Mediterranean and, from there, to the Italian cities, but about real diplomatic missions organized by the Papacy and entrusted to religious men of great physical and religious temperament.

The first mission, led by Brother Lawrence of Portugal, carried a letter "To the king and people of the Tartars" and was almost exclusively religious in character. He left Lyons in March 1245, but never reached his destination, his trail mysteriously lost in the Middle East.

A second mission was then set up, entrusted to six Dominicans: this time, the prelates succeeded in arriving in Persia and making contact with the Mongols, but the outcome was equally disastrous, because the threats of infernal punishment addressed to the Mongol leaders and other examples of a lack of tact on the part of the six clergymen immediately made the atmosphere combustible and thwarted even the slightest possibility of harmonious interaction.

In March of that same year, another small delegation led by a friar of the Order of Friars Minor of St. Francis, Giovanni da Pian del Carpine, left for the territory in Russia and Poland occupied by the Mongols. The friar carried a letter of a diplomatic nature, although the pope hoped in his heart that the mission would serve to convert the Great Khan and all his people to Christianity.

Giovanni da Pian del Carpine was born near Perugia, a few kilometres from Assisi. He knew St. Francis personally (he had, in fact, been one of the first followers), and thus seemed the right man for that assignment. Together with another brother, he crossed the endless steppes of Asia, the high mountain passes full of snow, the terrifying heat of the Gobi desert and, finally, he managed to make contact with the Golden Horde of Tartars. He stayed with them for two years and, upon his return, the account of his journey was revealed in the Historia Mongalorum, a rich ethnographic text in which the Franciscan friar described in detail the Mongols, their features, customs and habits.

Another important account of a religious and diplomatic mission is that of William of Rubruck, a missionary of the Order of Friars Minor, who, with his Itinerarium, left us one of the masterpieces of medieval travel literature. In contrast to the previous missions in which the Pope was involved, in 1253, Brother William was sent to the distant lands of the East by the King of France, Louis IX, with the task of meeting the Mongol leader Sartaq, since word had reached the West of his recent conversion to Christianity.

In spite of the enormous historical importance of the Historia Mongalorum and the Itinerarium, the fact that they were respectively written for the pope and the king of France meant that the diffusion and notoriety of these texts remained largely limited to curial entourages and court circles. They were obviously highly detailed accounts of the remote peoples of Asia, but, unfortunately, both books were undermined by the appearance of an underlying prejudice towards what was considered a community of barbarians who had not yet received the light of Christian truth. The authors had left Europe already ill-disposed towards the Mongolian peoples, and their encounters with them had done nothing but confirm all of their fears and prejudices, a fact that emerged clearly from the pages of their writings.

With Marco Polo, it was exactly the opposite: first of all, his tales became universally known throughout Europe, both in courtly circles and in the popular squares; moreover, they were full of positive energy towards those lands and peoples with their different approach to life, an approach that Polo not only respected, but also admired.

Marco was curious about everything, about the cities, yes, but, above all, about the history of the people who lived there, their way of life, their religion. The naiveté of his twenties played in his favour, shaped later by the experience of the great traveller and the enormous wealth of knowledge he had acquired on the roads to the East, among the mountains, the forests and the fearsome deserts. In short, with Marco and the extraordinary story of his travels, Asia became a land of happy wonders.

The Art of Travelling

It is useful to take a very brief look at the charts of *Il Milione*. We will do so by relying on the words of Marco Polo himself, or rather of the communication of those words by Rustichello da Pisa. Indeed, the game of the two narrating voices runs throughout the book: Marco's is the voice of the one who has directly seen things or heard them, while Rustichello has the job of making those memories interesting.

Here too, as in the *Commedia*, the protagonist of the story is the narrator himself, although, this time, he is assisted by an astute ghost-writer capable of weighing the words and the tone with which they should be put forward. This is immediately clear from the book's prologue, written in pure Rustichello style:

> *You will find here the narration of the immense and disparate marvels of the vast lands of the East—Great Armenia, Persia, Tartary, India—and of many other countries. This is an orderly and clear account that we have transcribed: and it was dictated by Mr. Marco Polo known as Milione, a wise and noble citizen of Venice who saw everything with his own eyes. And even if he did not see everything with his own eyes, he always availed himself of the testimonies of men worthy of faith.*

The few times in which the narration is in the first person resulted from Rustichello's experiences at court, which had taught him that it was much more incisive to state, to specify, to recount facts and things seen from close up with the tone of someone who was there. It is because of this overlapping of voices and their consistent interweaving that the text often oscillates between biography and autobiography, in the same way that it oscillates between reality and fantasy.

The text is divided into 183 chapters, few of which exceed two pages, with many being no longer than a few lines long: it is therefore a singular book, capable of narrating and combining the magical, the monstrous, and the unreal with the practicality and solidity of a merchant who is not afraid of the oddities that he encounters on his journey through the heart of Asia.

The first chapters provide a de facto flashback to the whole story, placing initial attention on the previous trades and voyages of the brothers Niccolò and Matteo Polo. While Baldwin was Emperor of Constantinople, around the year 1260, the two brothers had left Venice to try to expand their trade in the East, deciding to go beyond the Black Sea: "They bought a quantity of beautiful gems and left for Soldaia and from there reached the lands of Barca Kan, Tartar king who resided in Bolgara and Sarai on the Volga."

Unable to turn back because of a war that had broken out among the Tartars of the Levant, the two brothers decided to continue further along the road to the East. They travelled for long stretches through the desert, crossing vast windswept plains and following the mule tracks through snow-covered mountain passes. They finally reached Bucara, where the decisive encounter that would spur on the rest of the story occurred:

> *The two brothers were still in Bucara when a messenger arrived in the city, sent by Alau to the Great Khan. As soon as the messenger saw Niccolò and Matteo, he said, "The Great Khan has never met any Latin men, and he has a great desire to see them: if you want, I will take you to him. So they set out and walked for a whole year. The Polos arrived at the Great Khan.*
> *who welcomed them with great joy and celebration.*

The Great Khan proved very curious about the world from which the two Latins came, questioning them continuously. Eventually, he had the idea to send them as his ambassadors to the pope, with the request that the latter send him a hundred men, experts in the Christian religion, so that they could show the idolaters how that religion was superior to the others. The Great Khan gave the Polos a gold plate, and by presenting this plate, they were to be provided with lodging, horses, and escorts. The Polos took their leave and set out.

Back in Venice after so many years of absence, Niccolò was surprised to learn that his wife had died and that his son Marco, whom he had never seen before, had grown up healthy and strong, living with the family of another of his brothers. When they subsequently left for the Far East, the two expert merchants were accompanied by young Marco. The first stop on their journey was Acre, where they met the new pontiff, Gregory X:

> *The pope, judging that the request of the Great Khan was an event that would bring great good to Christianity, appointed two friars preachers so that they could exercise full religious power in those distant lands. Having received the blessing, they all departed, and with them Marco, son of Niccolò.*

However, the two preaching friars turned back hastily at the first encounter with marauders on the roads of the Middle East, not far from the coast, while the three Venetian travellers continued undaunted on their way. They crossed Persia, tried to embark for China from the port of Ormuz, changed their minds at the sight of the fragile junks used by the Arab sailors, turned back and crossed the mountain passes of the Pamir; then, they headed for Samarkand and entered Great Turkey, descended towards the Tarim basin and reached Tangut on the borders of Catai, ancient China. During this journey, which lasted three and a half years, they were regaled with all of the stories and legends that medieval Asia was steeped in, from that of the three magi to the clan of assassins, from the greedy caliph to the one-eyed cobbler capable of moving mountains, from the colourful world of the caravanserais to the camps of the Mongols. They travelled along the magical Silk Road and finally reached the court of Kublai Khan:

> When they came before the Great Khan, they knelt down and handed him the pope's papers and letters and the holy oil from the Jerusalem lamp. And when the Great Lord noticed the presence of young Marco, Niccolò said, "Lord, he is my son and your servant."

Marco, together with his father and uncle, remained at the court of Kublai Khan for seventeen years, becoming one of his most valued advisors. As the Great Khan's emissary, Marco Polo travelled to the four corners of his domains and, upon his return, the emperor of the Mongols was enchanted to hear the Venetian's tales of the great spice and silk trade he had seen in the Hubei valley, the sweetness of the gardens that had welcomed him in Yunnan, the superb stone buildings he had encountered on his way to Yangzhou, the waterways and fishing villages in the Myanmar peninsula, the beauty of the Turkestan women and the precious gems that embellished their necks.

In the course of his travels, Marco observed everything with great attention and without any prejudice: the skill of the craftsmen in softening the wood of the arches; the construction of water clocks; the practice of offering women to travellers passing through Kamul; the scaramantic and shamanic rites widespread among both peasants and court dignitaries. He also had the opportunity to observe the Khan's extended family and court customs up close:

> *The Great Lord has four wives, and considers them all legitimate. The eldest son who is born to one of the four wives is considered the rightful heir of the Empire on the death of the Great Khan; and by the other four the Great Khan has had twenty-two sons. The wives are called empresses, and each has her own court.*

He noted that, when not engaged in warlike operations or in receiving and talking with the emissaries of that Empire as vast as the Ocean, the Great Khan loved to go hunting:

> *In the place called Cacciar Modun found the pavilions of his sons, his women, and his knights. And here he holds court, and it is so large that a thousand knights can stand there. During the whole period of his stay he goes birding on the banks of lakes and streams hunting cranes, swans and other birds.*

For three months of the year, the Great Khan lived at Cambaluc, where he had a palace surrounded by crenelated square walls a mile long on each side, white and imposing: "On the four sides of the surrounding wall is a great palace rich and beautiful where all kinds of objects belonging to the Great Lord are kept."

The first of the year was celebrated in February with a luxurious feast, for which the Great Khan ordered all of his subjects to dress in white, the colour of happiness and contentment. Everyone gathered in the immense hall and offered beautiful gifts to the Great Lord. Some went so far as to give him a hundred thousand white horses.

Kublai Khan had also established, in Cambaluc, the Mint of the Empire, in which paper banknotes were printed, a practice then unknown in the West, where gold coins were used for trade instead. To print paper money, the Mint workers took mulberry bark, mixed it with glue and made a kind of bambagina paper out of it. Finally, once it was dried, they cut it into small rectangular sheets, longer than it was wide, and on each one, they affixed the seal of the Great Khan: "The Lord makes this coin be spent in every province. And no one dares refuse for fear of losing his life."

This was not the only thing that fascinated our Venetian. Another gem of the Mongol administration was, in fact, the postal system, entrusted to the messengers of the Great Khan: a messenger departing from Cambaluc would find a supply post every twenty-five miles in whatever direction he was going, where lodging and a change of horse were always ready for him. Between one resting-place and another, there were couriers who would run on foot, and, after three miles, find other couriers to whom they could pass the message, so that "the Great Lord has news from countries ten days' journey away in a day and a night."

And then there were the cities, for example, Cambaluc and Chinsai:

> *Leaving the town of Cambaluc, there is a large river called Pulisanghin, which flows into the Ocean Sea. A stone bridge, the most beautiful in the world, crosses it. It is no less than three hundred steps long, and ten people on horseback can cross it facing each other. It is all of gray marble magnificently worked with interlocking*

and has twenty-four arches and twenty-three pillars that support it rising from the water. [...].

In the province of Mangi is the noble city of Chinsai, "the city of heaven." The city is divided by an immense river, from which many canals branch off. The river has twelve thousand points of stone; it leaves the city and ultimately flows into the Ocean Sea. There are ten principal squares in the city, where forty or fifty thousand people come to market to buy all kinds of goods. There are many beautiful houses in Chinsai, well built and richly ornamented.

Like other Mongol leaders before him, Kublai Khan's life was marked by the sound of weapons and a boundless thirst for conquest:

Cipangu is an immense island in the middle of the Ocean, one thousand five hundred miles from the mainland. In the Cipangu there is gold in exceptional quantities, and no one takes any out of the island. You must know that the Great Khan, hearing of the great wealth of the island, decided to conquer it and sent two barons there. They took to the sea and sailed to the Cipangu, where they went ashore and conquered some lowland places.

But when they tried the same thing with Japan, it definitely didn't go as planned:

It happened one day that the north wind blew fiercely and the Tartars realized that all their ships were about to be broken. They immediately got into their ships and sailed out of the islands. They had sailed only four miles, when, by the force of the wind, their ships collided and were wrecked. Those who swam to a nearby island were saved, but those who didn't were all lost.

Things went better for Kublai Khan with the conquest of the kingdom of Mien and Bengal:

When the king of Mien and Bengal heard that the soldiers of the Great Khan had come to Vocian, he resolved to rush upon them with a great number of men. He gathered together two thousand elephants, and on each of them he placed a small wooden castle full of armed soldiers. The Tartars showed no sign of fear. They put their bows in their hands, drew their arrows and shot them at the elephants. They shot a lot of arrows, and the elephants were riddled with wounds. And so the battle went on.

Everything, however, inevitably comes to an end: so, after seventeen years spent in China and in the various provinces of the immense Mongol Empire, even such enterprising travellers and hardened merchants as the Polos were likely to experience that most common of emotions, homesickness. And to the melancholy was added another, much more important reason for wanting to return quickly to Venice: the advanced age of Kublai Khan and the inevitable uncertainty that would arise after his death. For the Polos, the opportunity to leave came when they were required to escort the young and beautiful Tartar princess Cocacin, betrothed to Argon, the Mongol governor of Persia, on a sea voyage.

Initially reluctant to let them leave, Kublai Khan finally relented and, as he had done twenty years earlier, gave the Polos passes in gold that guaranteed welcome in every part of his vast kingdom. They set sail from Zaitun, a port on the southern coast of China, their itinerary following the spice route. They docked in Java, from where they went on to the extreme point of the Strait of Malacca. They subsequently made stops on the coasts of Sumatra, the island of Ceylon and, after crossing the Indian Ocean, in Madagascar:

> *Mogdasio is an island towards the south of Middle India. In all the world not so many ivory tusks are sold as on this island. Know that griffon vultures live here, and those who have seen them say that they resemble eagles. The griffon is so large and vigorous that it can seize an elephant, lift it in the air to a great height, and let it fall to the ground. When the elephant has thus fallen, the griffin descends upon it and eats it and its flesh.*

After eighteen months of navigation, they finally disembarked in Ormuz, in the Persian Gulf. In Marco Polo's commentaries, this very long voyage is generally dismissed with meagre and lapidary remarks, but it doesn't take much insight to understand that it was actually a difficult and troubled journey, not to mention disastrous: as we have seen, of the 600 men on board (in addition to the sailors), only 18 survived.

At Ormuz, they learned of the death of Argon Khan, the Mongol governor to whom the princess was destined as a bride, but Cocacin was nevertheless welcomed and later taken in marriage by his son Casan. To recover from the fatigues of the journey, the Polos stayed for a long period of nine months at the court of Tauris, after which they set out again. At first, they travelled on horseback along the roads of Persia and the Middle East, until they reached Constantinople, and from there, they embarked for Venice, where, twenty-four years after their departure, they disembarked in 1295.

The Never-Ending Querelle

Since its publication, *Il Milione* has been accompanied by doubts about the authenticity of the journey made by its author. Even Marco Polo's closest friends were so consumed with suspicion that, on his deathbed, they begged him to tell them the truth: his response was that he had not told them even half of what he had seen.

Scores of sinologists and scholars of Mongol history have gone in search of the princely evidence that would prove, once and for all, that the Venetian never made it beyond the Persian ports and that, in fact, he had learned of the things that he narrated from the accounts of Arab sailors and traders.

The loudest voice of the sceptics was that of Frances Wood, director of the British Library's China collection and author of the book *Did Marco Polo Go to China?* According to Wood, the Venetian merchant's journey appears to be completely inconsistent, both geographically and chronologically. The British sinologist also pointed her finger at the fact that the names used by the author are neither Chinese nor Mongolian, but Persian, which seems to validate the hypothesis that Polo never went further than Persia or the Black Sea.

Wood also pointed out that, strangely enough, some famous Chinese customs are never mentioned in *Il Milione*, such as the art of tea, the use of chopsticks, the binding of women's feet, the writing of ideograms, the printing of books or fishing with cormorants.

Another serious indication of the lie perpetrated by Marco Polo, according to the British sinologist, is the absence in the text of any mention of the Great Wall. The Venetian traveller also claims in his book to have been governor of Hangzhou for three years, but his name has never been found in the official archives of the city. Further indications of guilt would be the pages dedicated to the participation of his father Niccolò and uncle Matteo in the Mongol siege of Xiangyang, an important stronghold of the Song dynasty in Hubei: in fact, it is easy to demonstrate that if the Polo family ever reached China, they certainly would only have arrived the year after this siege.

All of these doubts have been addressed and resolved by an equally vast array of scholars and historians of China. First of all, many of the omissions attributed to Marco Polo can also be found in accounts by other travellers of the time, such as the Arab Ibn Battuta. Moreover, unlike Dante Alighieri, Marco Polo was certainly not a writer, but a merchant: even if he showed himself to be an acute observer of world affairs, he had always had a very rudimentary scholastic education, lacking elementary notions of rhetoric and literature. There are very few personal notes in the book; what matters most to him are, in fact, the descriptions of places and peoples, among other things, drawn up many years after his return home and while in captivity in the Genoese prisons, and thus inevitably subject to error for a simple lack of memory.

Several scholars have also pointed out that, during his long stay in China, Marco never actually mixed with the Chinese, for a very simple reason: the Chinese were a population subjugated by the Mongols, who treated them as slaves. At the court of Kublai Khan, there were instead a large number of Persian and Turkish merchants, and the lingua franca of the time in almost all of Asia was Persian. Chinese was the language of the dominated and Mongolian that of the dominators: between the two communities, there was an abyss of incommunicability, broken only by the few sporadic contacts of an economic and administrative nature. Paradoxical as it may seem, there is reason to believe that the lack of mention of some of the millenary Chinese customs is therefore due to the absence of contact, of any real exposure of Marco to that population. Moreover, if the Venetian does not mention the Great Wall, it is most likely because, at the time, that fortification was a discontinuous set of walls made of mud and propped up with wooden poles, in simple terms, an engineering work that could hardly have impressed our traveller.

On the other hand, it should be pointed out that Marco Polo has left us an extraordinarily detailed description of Cambaluc and other cities in northern China, a description that would have been difficult to write if he had not actually been present in those places for so long.

How, then, can we explain the false histories about the siege of Xiangyang and the position of governor exercised in Hangzhou? According to the scholar Igor de Rachewiltz, there is no reason to doubt that Marco Polo lived for a long time in Hangzhou, a city in many ways similar to Venice, perhaps carrying out a high office assigned to him by Kublai Khan, although not that of governor.

In this case, we are most likely in the presence of a vacuous exercise in vainglory, with the simple intent to embellish the story and magnify the role he played as much as possible: in this regard, we should also take into account the different sensibility of the modern writer compared to that of the medieval writer, for whom it was good practice to embellish the text with surprising, unreal and fantastic details, sometimes clearly false, and perhaps inflate his role to "create a tone."

For that matter, the tales of Odorico da Pordenone and Giovanni da Pian del Carpine also contain a good deal of fantasy, and they were both religious… Why would we expect something different in *Il Milione*, when the plot and the embellishment of the text were the prerogatives of Rustichello da Pisa, a writer of court and chivalric texts?

Venice and Florence
In Which We Learn of St. Mark's Horses, of Byzantine Elections, of a Cursed Flower, of a Poet's Impatience, of Darkness and Deception

At the Dawn of the Thirteenth Century

The thirteenth century opened with a great dream of universal empire. This idea, which, in the past, had driven Alexander the Great to the insane enterprise of dominating the deserts of Africa and reaching as far as the gorges of Persia, and which had encouraged the Roman generals to conquer every remote corner of the then-known world, never seemed so close to realization as in the thirteenth century, both in the West and in the East. For while, in the heart of Europe, the foundations of a majestic papal theocracy were increasingly being consolidated, with kings, emperors and princes gravitating around it, in the distant prairies of deep Asia, ferocious warriors raged on horseback, armed with bows, arrows and long curved swords, flat-faced men with almond-shaped eyes: they were the fearsome knights upon which the largest empire that had ever existed, the Mongol Empire, was founded.

In spite, however, of these encouraging political premises, the realization of a universal Empire turned out to be a mere utopia. The Christian Empire, caught in the grip of the fearsome Islamic threat at its borders, was, in fact, disintegrating under the centrifugal thrust of all of its countless components and because of the fight without quarter between the Papacy and the Empire. The Mongol Empire instead collapsed under its own weight: it was simply too vast and heterogeneous, its subjects being the Chinese, the Turcomans, the Kirghis, the Tungusi, the Persians and the many trickles of ancient peoples snaking across of the northern steppes. It was an Empire that had been built too quickly to last long after the death of its founder, the ruthless Genghis Khan. The swan song of the Mongol Empire came with Kublai Khan, Genghis Khan's grandson, the last great solitary emperor of the vast Eastern territories.

Two exceptional witnesses of that troubled era, full of political, military and economic upheavals, as well as pregnant with new ideas in the fields of philosophy, science and geography, were our own Marco Polo and Dante Alighieri, the Merchant and the Poet.

The two were almost the same age, only eleven years apart: the first was born in Venice on September 15, 1254, the second in Florence under the sign of Gemini, as Dante himself says in *Paradise*, most likely on an unspecified day around the end of May 1265. While the exact date of the Poet's birth is still unknown, the date of his baptism is, having taken place on March 26, 1266, a Holy Saturday, in the Baptistery of San Giovanni. The name with which he was baptized was Durante, even if he was always known by the diminutive Dante.

Lauders and Detractors

Boccaccio was a passionate admirer of Dante and an acute commentator on his work, which he got to know in his earliest youth and which was a source of inspiration for him: in the *Corbaccio*, for example, the presence of the forest and the vision unequivocally refer to the infernal setting of the *Commedia*; one of the tales in the *Decameron* is dedicated to Ciacco, the protagonist of Canto VI of the *Inferno*, known for his gluttony, and Filippo Argenti, another famous Florentine portrayed by Dante, also appears. Boccaccio wrote *Trattatello in laude of Dante*, which is also a hagiographic biography, in which true facts are mixed with miraculous events, but for which he sought documents and testimonies among the circle of people who had known the Poet, including his daughter Antonia/suor Beatrice in Ravenna. In addition to having copied three codices of the *Commedia* in his own hand, Boccaccio also gave lectiones magistrales on the *Canti dell'Inferno*, and was also responsible for adding the adjective "Divine" to the title of Dante's *Commedia*, as a seal of his admiration for the poem.

Francesco Petrarca's attitude was completely different, as he was always openly impatient with and critical of Dante, probably driven by envy, given the great popularity enjoyed by the Florentine poet. On the other hand, the two were siderally distant on everything, on the conception of the world, on man's place within it, and on the role of women and literature. To best express his aversion to Alighieri, Petrarch played the card of iciness: in reply to a letter from his friend Boccaccio, who had sent him a copy of the *Commedia* as a gift and had invited him to read with sympathy and appreciation the verses of someone who, in life, instead of honours, had suffered the damnation of exile, Petrarch replied that he could only feel contempt for a man whom he had not thought about since he was a child. He added that his preference was for Latin, as poems in the vernacular were mostly recited by ignorant persons who defaced them, and he did not wish to run this risk; moreover, he had never wished to study the writings of Dante because he wanted to create his own style, without being influenced by any great master. And on this, perhaps, he had a point.

Boccaccio, a passionate and unswerving devotee of everything that had to do with his famous predecessor, gives us, in his *Trattatello in laude di Dante*, a version of the birth of the Poet full of evocative echoes of the past, even portraying the event as having been announced to the Poet's mother in a dream. She is lying under a large laurel tree, in a vast green meadow and near a spring of clear water: once she has given birth to her son, he feeds himself on berries from the laurel and drinks from the spring, suddenly taking on the appearance of a shepherd boy. When he tries to grasp the branches of the tree with his hands, that little shepherd falls ruinously to the ground, but, getting up again, he is transformed into a magnificent peacock ... Dreamlike visions and the desires of a mother, you might say, but they certainly point a finger towards the future man, the Dante who was haughty, proud and deeply convinced of his uniqueness, an attitude that would lead him to long, hard years of exile, as predicted by his ancestor Cacciaguida:

And you will taste the saltiness of bread
when offered by another's hand—as, too,
how hard it is to climb a stranger's stair.

Paradiso, XVII

In addition to the glitter of the peacock's feathers, there was also another resounding maternal omen, the appearance of a comet during the months of her pregnancy "with great rays and hair behind." Everything—the stars, divine design, the moods of the cosmos—seemed to conspire to announce the arrival of a child destined to be marked as a watershed in the history of poetry, philosophy and politics. Florence, truth be told, had always been a haughty and self-important city, and therefore there was no reason to expect that one of its most illustrious sons would fall short in these particular attributes. And the Venetians, upon the birth of Marco Polo, were certainly no less proud than the Florentines (Venice was considered "the noblest and most beautiful city"), with the whole city exulting: "A lord of the world is born." What they could not have known was that, with the discovery of the Orient and distant China, the opening of new routes and boundless horizons, the explosion of trade and the sweet scent of spices, Marco would not only be a hero of that new fourteenth-century world, but also a dedicated explorer and primary chronicler of it.

The World Seen from Rialto

On July 23, 1268 a new doge, Lorenzo Tiepolo, was elected in Venice. The celebrations were spectacular and enjoyed the enthusiastic support of the merchants and all of the artisan guilds. Martino da Canale, one of the first historians of Venice, was also present, and he has left us a vivid account of the euphoria of those days: the first to parade down the Grand Canal, in front of St. Mark's Square, overlooked by the Doge's Palace, was, needless to say, the extremely powerful Venetian fleet, the creator of the predominance exercised by the Serenissima over all known seas, the Adriatic, the lower Mediterranean, the Aegean, the Dalmatian and African coasts, the ports of Crete and Malta, Constantinople… From the decks of the fifty galleys and the other ships that accompanied them, there were loud shouts of jubilation, the roll of drums, and the imperious ringing of bells on the bow.

Then, it was the turn of the interminable crowd of revellers on foot, who, accompanied by musical bands, advanced singing and shouting: "Long live our lord, the noble doge Lorenzo Tiepolo!" Then followed the weavers, wrapped in their gaudy silk cloths edged with gold; then, the iron smiths, with their coloured hats and their flags; the glassmakers, with their rich pearl jewels, holding precious glass trays ornately decorated; then, the goldsmiths, covered from head to foot with the most precious of metals, their hands adorned with rings of sapphires, emeralds and rubies, and around their necks topazes, hyacinths and amethysts, all stones that came from the far East. Each guild entered the square from the narrow calli, passed under the majestic quadriga of the cathedral, and then paraded in front of the newly elected doge, standing on the balcony of the Doge's Palace, to greet him with shouts of jubilation.

Among the thousands of people who attended that great feast and that display of opulence was a young man of fourteen, with a quick step and a lively intelligence, his eyes full of curiosity about all things in the world: Marco Polo. He too was inebriated by that rainbow of colours, the voices, the notes of the flutes, the silver cymbals and the low timbres of the trumpets, the glitter of the flags and the numerous banners that showed that Venice was the queen of the seas, the undisputed mistress of all maritime commerce.

Since the age of ten, Marco had gotten into the habit of exploring the hidden corners of that magical city, of discovering the secret passages that connected the campielli. He knew the scent of the gardens behind the churches, the beautiful little islands of the lagoon, Poveglia, Torcello, San Giorgio. He also spent many hours a day sitting on the banks of the Zattere, along the quays of the port where, with a dreamy gaze, he would lose himself behind the waves of the trabaccoli (lugger boats) from Istria loaded with salt and the barges full of timber that arrived in the lagoon by sailing up the Po delta, but, above all, behind the wake of the ships returning from Alexandria in Egypt, from Acre—the gateway to Palestine -, from Rhodes, Malta, Zara and Constantinople.

The enormous Latin sails of those galleys entered the Grand Canal and headed decisively for the Rialto, the nerve centre of all of the commercial exchanges that took place in the city, of the negotiations and bargaining that became more and more frenetic every day: in the "business square" of the

Rialto, prices were established, new commercial enterprises were promoted, spices, camphor and incense were stocked. Additionally, there was ivory, perfumes and textiles, carpets, pearls and the highly prized silk, which, on the backs of camels, had crossed the desolate highlands of Asia to reach the Caspian Sea and then Venice.

Slaves were also offloaded at the Rialto from the most unthinkable places in the world, adding a further note of exoticism to the babel of voices of the Florentine, Jewish, Greek, Macedonian, Slav, Levantine, Syrian and Armenian merchants who thronged those quays and crowded the taverns and inns around them. To the noises of the merchants were added those of the shipowners, of the sailors in search of boarding, of the caravaners who, their horses laden with skins and fine wool, were ready to depart for France, Germany and Flanders. Seen from the Rialto, the world undoubtedly looked like a great bazaar: the heart of all commercial opportunities and the epicentre of fortunes and riches were there, in Venice.

Young Marco always asked the sailors that he met for news of his father Niccolò and his uncle Matteo, who had left Venice years before for Constantinople, dreaming of a great trade with the East; They had then moved on to Soldaia, in the Crimea, where they had a warehouse, only to disappear, swallowed up in the immense deserts of the Tartars, in that unknown land of which only fragmentary news about fierce warriors, powerful shamans, terrible killers, beasts with big horns and salamanders able to walk through fire arrived, in other words, all of the folklore of war, magic and madness that has come to characterise our Middle Ages.

Marco knew neither his father's face nor the colour of his eyes. He had never heard his voice or felt the warmth of his hand on his cheek. Niccolò had left months before his son had even come into the world. It was only thanks to his mother's stories that he knew that this man was not afraid of anything, that he would return one day to Venice, but would likely leave again the next day, always moved by the frenzy of trade and travel.

However, his mother had been dead for several years and he had been taken into his uncle's house, in the Dorsoduro district, facing the Giudecca and Grand Canals.

Marco had grown up in a world dominated by commerce and, as the son and grandson of merchants, he had immediately learned the art of counting, of weights and measures, the squiggles of the alphabet and the richness of the Venetian language: he always listened with great attention to the tales of the fabulous journeys of those who returned from distant lands, from the coasts of Gibraltar or the hills of Palestine. He dreamed of one day going to the farthest place on earth, in the company of his father: he knew the way, Marco was certain; it was enough to rely on the sun or on the needle of a compass.

An Extraordinary Republic

Geology is the science of long time, those breaths of the Earth that shape coasts and the course of rivers. Thousands of years, infinite tides and the uninterrupted flow of that which rivers such as the Po, the Adige and the Piave pour into the sea had created lagoons and fens, pockets of sand and marshes in the northern Adriatic, all dotted with small islands full of shrubs and water birds. Those islands of the Venetian lagoon, those muddy archipelagos proved to be of good fortune for those who lived inland: when the first waves of terrible barbarians arrived, especially Attila's Huns, many Venetians fled their homes, preferring to take refuge on those unhealthy, damp and muddy stretches of land rather than face the trail of death and destruction that accompanied the hordes of vandals.

Once settled on the islands, the lagoon dwellers began to work industriously, first with the art of fishing, then with that of salt, and finally with the art of commerce. As long as they limited themselves to trading a few goods just outside the lagoon, there were no problems with the Byzantine emperor, who, although his heart was in Constantinople, had moved his court to Ravenna, enriching it with a hundred churches and a thousand mosaics. For the Byzantine merchants who travelled between East and West, the emporium of Torcello or the market of Rialto simply became one more resource at their disposal: their real rivals were the Arab merchants, who unscrupulously loaded their merchandise into the bazaars and emporiums on the coasts of Syria and North Africa and in the ports of Spain.

> **The Spice Journey**
>
> The opulence of Venice was also well represented by the infinite number of spices with inebriating perfumes that came from distant and unknown lands. In order to reach the ports of Syria and Egypt, the places from which the Venetians brought them to the shores of the Rialto, the sacks of pepper, cloves,

cinnamon, turmeric, nutmeg and dozens of other spices and aromatic herbs made a very long journey by sea and land. From the distant lands of India, Malaysia and China, they were crammed onto the feluccas of the Arab sailors, who, going up the Red Sea, unloaded them in Jeddah, the port of Mecca. From there, on the back of camels, the bales of these precious goods would cross the desolate deserts of the Arabian peninsula before finally reaching the docks of Damascus and Alexandria. The various passages were regulated by dozens of intermediaries, all to be paid, of course. However, the Venetians soon became unsurpassed masters of that subtle and cunning game that is the art of commerce, slowly replacing the Byzantine merchants who had first opened the doors to the god of money, managing to become major commercial players even in those merchants' very own hometown, in the heart of Constantinople!

All of this changed around the year 1000, when the Venetians became aware of their innate sense for business and boldly began to sail towards the Levant, towards that immense source of earnings offered by the new markets. They had discovered that one could buy and resell anything: carpets and slaves came from Alexandria in Egypt, gold came from the shops of Asia Minor, precious stones were bought in the emporiums of Corinth and resold in those of Corfu, and then there was timber, alum, cotton, silk, the vast array of all armaments. There seemed to be no limit to Providence....

All of that wealth came from the waters. The origin of one of the oldest Venetian festivals, the historic "Marriage to the Sea," dates back to those years when the Doge, coming out into the lagoon on the Bucintoro, with a large procession of boats of all shapes and colours, would, after a blessing from the patriarch, throw a gold ring into the water, pronouncing in Latin the famous formula: "We marry you, O sea, in token of eternal dominion."

A Byzantine Election

The position of doge was of Byzantine origin, therefore, it required an equally Byzantine election... It was a real lottery, a strange cabal based on the pure laws of chance. The fundamental steps are worth noting: once the assembly of the Maggior Consiglio had met, the youngest councillor had to leave the Doge's Palace and take the hand of the first child he met on the street. The child, from then on called a balotin, was charged with extracting balls from an urn: there were as many balls as there were possible voters, but only on thirty of these balls was the word 'elector' written. The balotin would extract a ball for each voter, but only those who had received the ball with the magic word on it would meet again to draw nine more of them. The nine thus drawn would meet and chose another forty councillors (the first four would propose five, while the others would propose four each). The latter, by means of a new election, would select twelve, who, in their turn, would elect another twenty-five persons, who would further choose, through the drawing of lots, nine of their own number. The newly selected nine would then elect another forty-five councillors, from whom eleven would be drawn by lot: from these would

> be elected the forty-one "great electors," who would then actually choose the Doge. Ultimately, the highest office of the Republic went to the person who obtained the most preferences, with a minimum of twenty-five votes. This overall mechanism, needless to say, was maximally disruptive, in a good way, preventing any attempt to plan the final result, which was therefore a pure result of chance. In Venice, there had never been a feudal nobility that opposed the merchants: the merchants themselves were the nobility of the city, the undisputed masters of all the wealth that arrived every day from a dense capillary network that extended to the borders of the known world and that flooded the quays of the riva degli Schiavon and the Rialto square.

The Arsenal of Venice became the beating heart of the Venetian naval industry: from its slips, the *arsenalotti* put *palandre* and *dromoni* into the water, great galleys of war and commerce, and the fearsome *brulotti* with which to destroy enemy ships. A large portion of these merchant ships were in public hands, just as the regular commercial expeditions were, the "muda," in which private individuals participated by renting space in the holds to transport goods. In the space of a few years, with brutal and constant aggression, Venice expanded the boundaries of its commercial universe beyond measure: the skill and cynicism of its ambassadors, together with the power of its military fleet, became the premises for the irresistible rise of the Maritime Republic. *Il Fontaco dei Turchi, il Fontaco dei Tedeschi* and the dozens of other "fondaci" scattered around the Mediterranean were the basis for the Republic's irresistible rise, always overflowing with goods; Venetian merchants were everywhere, and their insolence became legendary, with "Venetian" ending up being synonymous with an unscrupulous, authoritarian man, a ruthless person, as Shakespeare would have said.

The extraordinary Venetian Maritime Republic was all about favouring wealth to the maximum, a very well-oiled oligarchy having been set up for the purpose: the extension of trade went hand in hand with the extension of all sorts of privileges, to ensure that power remained concentrated within a small number of families, thus excluding the people from any elective function, but, at the same time, trying to limit the power of the doge, the supreme office of the Serenissima.

The Horses of San Marco

If the year 1000 had been the turning point of the commercial fortune of Venice, in the thirteenth century, things went even better: for a long time, Venice had inserted itself into the great traffics of the crusades: landing on the coasts of Palestine and unleashing hordes of Templars, Teutonic knights or simple armies of beggars had proved to be an extremely profitable initiative. It had also provided the opportunity to build a real maritime empire, as they were able to rout all of the other ports along the Adriatic, such as Zara and Ragusa.

The occasion of the grand slam, the bargain of a lifetime, finally came with the fourth crusade, perhaps the most ramshackle, because it was not organised by kings or emperors, but by an ill-assorted group of feudal lords from all over: they presented themselves before Doge Enrico Dandolo to negotiate the transport of a great mass of infantrymen, squires and knights to the ports of Haifa and Acre, along with sufficient food supplies for a year. The price agreed upon for the operation rose to the exorbitant sum of 85,000 marks of silver, equal to more than twice the annual income of the King of France.

The old doge, a shrewd politician and merchant, was well aware, however, that the crusaders had overestimated their financial capacities, but, precisely for this reason, he had given them his backing; in his eyes, this enormous debt was the necessary premise for the Republic to carry out the most lucrative enterprise in its history. As expected, in fact, at the moment of departure, things began to go wrong: the crusaders present on the quays of Venice turned out to be far lesser in number than expected and the money they had collected was not enough to cover the agreed-upon price. Thus began the first act of this drama, in which the Venetians played the part of the offended, refusing to put to sea if the pacts were not respected, that is, if payment of the entire negotiated sum was not made. But the matter took a bad turn at a certain point, with the Venetians's behavior getting out of hand: their mood poisoned by the stalemate of the negotiations and by the intolerance of that forced stop, those soldiers with bad intent began to create havoc in Venice, harassing women, getting drunk and coming to blows with the first person they came across. Camped on the Lido, they had transformed that beach and the dunes behind it into a filthy mess.

It was therefore necessary to find a solution as soon as possible, leading to Act Two of the play. In the meeting between the Doge and the leader of the crusaders, Bonifacio I del Monferrato, an unscrupulous agreement was finally reached: the Venetians would become crusaders themselves, on condition that they could pocket the booty obtained from the conquest of Zara, a

Dalmatian city that had dared to rebel against the dominion of the Serenissima. The agreement also provided that the army would be commanded by Doge Dandolo in person.

So, the imposing Venetian fleet finally left the lagoon to the shrill notes of the trumpets and the singing of the Veni Creator Spiritus, setting sail southwards. Only a very few aristocrats of the restricted senatorial oligarchy knew the true purpose of the expedition. The chronicler of that enterprise, Goffredo di Villehardouin, in his reports, underlined that never had such an imposing fleet been seen before, composed of more than 200 ships, with about 17,000 Venetian soldiers and 32,000 crusaders on board. It took no time at all for the galleys to arrive under the ramparts of Zara and besiege the city from the sea: a few days later, on November 15, 1202, the assault by land was ordered and the city was immediately conquered and sacked. With winter approaching, it was decided that they would remain in Zadar until the arrival of spring.

Pope Innocent III, however, was horrified to learn that a Christian city like Zara had been prey to the bloody sacking of the crusaders. An excommunication was issued for all those who had participated. The French and German barons at the head of the crusading troops justified their behavior by saying that they had been forced, under blackmail, by the Venetians to attack the Dalmatian city, and the pope thus forgave them. For his part, Doge Dandolo did not care about the papal bull and proceeded to execute a new plan: to turn the sails of the galleys towards Constantinople, with the excuse of restoring the legitimate holder of the imperial seat, which had been vacant for some time.

Once again, the imposing mass of people who had been idling for months in Zadar embarked. A few weeks of sailing were enough for the unmistakable silhouette of the Golden City to appear on the horizon, with the dome of St. Sophia, the hundreds of towers of the walls and the Grand Palace. The ships having been positioned under the ramparts of the old capital of the Eastern Empire, a cloud of arrows began to rain down on the wall towers. "Darken the sky with your arrows!" cried the doge. Then followed the final assault and, on July 17, 1203, the conquest of the city.

The Venetians freed the old Emperor Isaac II from prison. The latter, as the first act of the restored Empire, appointed his son Alexius Comnenus co-regent. In truth, it was Alessio himself who had contacted Dandolo and solicited his armed intervention: a real pact with the devil, as Alessio himself realised once the crusaders and the Venetians had become masters of the city and were bivouacking in the streets, waiting for the enormous sum promised for their services.

In order to meet this debt, the young emperor antagonized everyone, both the people, with the new and heavy taxes he had introduced, and the clergy, having had the silver candlesticks requisitioned by the churches melted down. The situation quickly began to degenerate, with skirmishes becoming increasingly frequent between the Crusaders and the Byzantines. The natural outcome of this explosive situation was a revolt, led by Alexis V, known as Murzuflo. Comnenus was captured and strangled, while his father, Isaac II, died shortly thereafter under mysterious circumstances. Once in power, Alexis V categorically refused to pay the Crusaders and the Venetians, and even ordered them to leave Constantinople. This was the beginning of the city's end.

The main drama took place on April 12, 1204, a few days before the "Easter of flowers." On that date, the valiant Christian knights gave vent to all of their greed and cruelty, devastating houses, stripping churches of their precious furnishings and the relics of saints, and carrying out bloodbaths. Fires broke out everywhere and there was savage plundering of everything that could be stolen, torn, or vandalized. The ancient Byzantium was put to fire and sword and plundered of every good; the doge Enrico Dandolo had no pity for his former allies. Among the innumerable works of art that he claimed as spoils of war were the four enormous horses in splendid gilded bronze that, having arrived in Venice, immediately found their place above the portal of St. Mark's Basilica. On his return to his homeland, the old doge was welcomed as a hero.

Once the raid was over, the rituals of politics and diplomacy took over. When Dandolo refused to become the new emperor of Constantinople, the Venetian and Crusader commanders agreed to elect Count Baldwin IX of Flanders. But Venetian influence expanded further: Dandolo claimed for Venice the west coast of Greece, the whole of the Peloponnese, Gallipoli, Adrianople and the ports of old Thrace on the Sea of Marmara. The Venetians also purchased the island of Crete from the new emperor and, as if that were not enough, they claimed three-eighths of the city of Constantinople, settling in a district behind the basilica of Santa Sofia.

All this went on until 1261, when the Byzantines signed a treaty of alliance with the Genoese, historical enemies of the Serenissima, by virtue of which, in exchange for help in the reconquest of Constantinople, the Genoese would then enjoy all of the privileges held by the Venetians since the beginning of the thirteenth century. Episodes of piracy and violence between the two maritime republics had always been the order of the day, but what was happening then seemed like a dangerous attack on the prerogatives of Venice, and the worst of nightmares materialized when Constantinople fell into the hands of the Genoese. The Venetians stationed in Byzantium were forced to flee or were killed, their wealth was requisitioned and their foundations burned. Anticipating the disastrous outcome of that war with the Genoese, the two Polo brothers, who had been present in the markets of Constantinople for some time, decided to leave that increasingly insecure city and go to Soldaia on the Black Sea, to try to open up new markets towards the East. Their future meeting with Kublai Khan threw open the doors to Asia.

The Art of Wool

If the internal political cohesion of Venice was the basis of its economic success in the markets of Europe and the ports of the Mediterranean, Florence was, contrastingly, torn for a long time by violent internal struggles, fuelled by hatred between ancient families, a hatred without limits in which the name of the protagonists changed every so often while the destructive force remained intact. Except for those directly involved, this was all incomprehensible to most people: the Venetian doge Andrea Dandolo, for example, thought that Guelphs and Ghibellines were the names of two brothers in Tuscany, even though the destructive power of that fratricidal struggle did not escape him at all.

As is well known, the terms "Guelphs" and "Ghibellines" were rough Italian translations of the names of two German houses of the twelfth century who fought each other for the crown after Emperor Henry V's death, which occurred in 1125: the Welfen, supported by the pope, and the Hohenstaufen of Swabia, lords of the castle of Wibeling. In Italy, the mispronounced names of these two families began to circulate on the occasion of the clash between the Papacy and Frederick II of Swabia, and soon ended up designating the pro-papal parties (the Guelphs) and those who sided instead with the emperor (the Ghibellines), although these ideological clashes, so noble and high in politics, were often only smokescreens for continuing fierce local rivalries, born out of much more miserable circumstances.

Those were the years when various factions began to impose themselves on different municipalities of the peninsula, arming families against each other, turning the houses of the powerful into fortalices and turning political opponents to arms, when they could, in order to occupy or destroy their enemies' properties. When these internal struggles became intertwined with foreign policy conflicts and broader ideological issues, the consequences were dramatic. This was the case in Florence during the period from the beginning of the thirteenth century to the late fourteenth century: Dante Alighieri was not only a direct observer of those fierce clashes, he also paid the bitter price of exile for his irrepressible passion for politics.

Life in Florence at the dawn of the fourteenth century appeared as tangled as an old olive tree trunk: all of its knots were actually much older in origin and dated back to the struggles undertaken by the city to free itself from the medieval bottlenecks that oppressed it. In fact, the conflicts dated back to the beginning of the thirteenth century, when all the unbearable political heaviness of the Middle Ages was plastically embodied by a long series of castles scattered throughout the hills around Florence, each with a local squire surrounded by his large family, families of high lineage, and various high-sounding names: Cadolingi, Aldobrandeschi, Frescobaldi, Uberti, Guidi, Donati, Adimari. These were overbearing people proud of their privileges—such as the collection of taxes, the usufruct of the crops or the levy for those who passed through their territories—and willing to do anything to hold on to them. Each one had his own personal army, easily recruited from the nearby villages: the countryside was full of people willing to fight.

Florence set up its first rudimentary city government only at the beginning of the twelfth century—about a century later than its neighbours Pisa and Lucca—and the first wars undertaken by the newly constituted Commune were directed against those Lords who surrounded the city. Each quarter of Firenze agreed to provide the army with its own company of soldiers, led by their own commander and gonfalon: at the roll of the drums, that variegated troop of crossbowmen, infantrymen, horsemen and feditori, dressed in their gaudy clothes and heavy armour, moved to conquer the various castles and, every time they conquered one, they would immediately raze it to the ground.

History repeated itself several times and, seeing how bad things were, many of the lords decided to compromise in order to save what could be saved: in exchange for Florence's recognition of their landed property, they promised to swear loyalty to the new-born Commune and not to interfere with its politics. Several of them opted to settle in the city. In other words, they urbanized, bringing with them a whole queue of relatives, soldiers and servants, but they were obviously not content to have an insignificant house of wood and straw like the vast majority of houses in the city at that time: with the arrival of the castellans in the city, solid masonry palaces were born, rising to the sky with towers that were up to eighty meters high, the categorical imperative being to have a tower higher than one's rivals.

> **A Forest of Towers**
>
> Florence in the thirteenth century was invaded by a forest of towers, hundreds of them, almost like an urban transposition of the towers and castle wings that had previously invaded the hills of Mugello, Val d'Arno, Calenzano and Campi Bisenzio. On the façade of the palaces, there were large rings for attaching horses, donkeys and mules, while the hooks installed on the first floor were used for hanging curtains or large drapes during solemn processions, receptions for foreign ambassadors or public festivals. Very often, near the main door, there were also metal rails where torches could be hung to light up the streets, which were often the scene of attacks at night. During the day, contrastingly, those same streets swarmed with people, who were around from the morning Angelus until the one in the evening. At the same time, it was in the streets that the activities of blacksmiths, cobblers, barbers, tailors and junk dealers took place; artists, painters and engravers also worked under the canopies of the loggias. The street corners were usually occupied by the municipal heralds, ready to read the latest ordinances. To this picturesque crowd of people were added the peasants, with their carts full of the fruits of the earth ready to be sold at the market, as well as a hearty troop of beggars and petty thieves.

Entire districts developed around these new magnate families, since the prestige of the family was judged not only by the height of the bell tower, but also by the extent of the shops, warehouses and houses gathered around the main complex. Everyone tried shamelessly to show off what was left of their original power. The greatest danger—to be avoided at all costs—was that other families might encroach upon their territory and control its accesses.

Thanks to the tumultuous development of the city, the initial 10,000 inhabitants of the twelfth century—the time when the crusader Cacciaguida, Dante's great-great-grandfather, lived—soon turned into 50,000 by the beginning of the thirteenth century, reaching the remarkable number of 100,000 by the beginning of the fourteenth century, in Dante's time. As a consequence of this growth, the old walls protecting the city had to be enlarged: at the time of Matilda of Canossa, the original Carolingian circle had already been enlarged to contain a second, much wider wall, and this was further enlarged in the years 1172–1175 until it stretched beyond the Arno and encompassed an area four times the original size.

As was easy to foresee, however, with the nobles no longer in the surrounding castles, but firmly established in the city, Florence found that its enemies were now at home. The ancient feudal lords had understood, in fact, that it was sufficient to control the Council of One Hundred from the outside to have full control of everything that happened inside and outside the city walls. The wealth of the countryside was enough for the magnates to buy favours, set up blackmail, manoeuvre behind the scenes and acquire great benefits.

All this, however, began to be intolerable for the "new people," who, since the first half of the thirteenth century, had become rich through trade and production, especially of wool. The Florentine wool workers had soon learned how to treat it and had revolutionized the manner of dressing throughout Europe. The woollen cloth, with its very fine yarns and short fibres, both in terms of weft and warp, proved to be a gold mine, an inexhaustible source of earnings, starting with religious clothing! The number of clerics to be clothed, in fact, was never-ending, whether they were the humble friars in the monasteries or the cardinal-portates of Rome. But the real rain of money began to arrive with the invention of coloured stockings, of fustian, of trousers that covered the legs instead of the sad, bare cotton shirts, and with the sale of tunics that were more and more elaborate and specially designed. Every powerful person (or anyone who thought he was powerful) was eager to have those well-combed woollen jackets that fell magnificently down the bust, the garnets that enhanced the body, with the puff of the cloth sleeves that opened in folds of a thousand colours, the capes with pointed hoods, the two-coloured brocatelle tunics tightened by large leather belts.

The Meeting with the Great-Great-Grandfather

There was always a lacerating love-hate relationship between Dante and Florence, and the *Divine Comedy* is a faithful testimony to this personal drama. A particularly significant encounter is the one that takes place in Paradise, in the sky above Mars, between Dante and his great-great-grandfather Cacciaguida, who unleashes an outburst of words of disdain for the corruption into which Florence has fallen and deeply regrets the purity of ancient customs.

The spirit introduces himself to Dante as the progenitor of his family: "I was your root." He was, in fact, the father of his great-grandfather Alighiero, whom Dante had previously met in Purgatory, where, for more than a hundred years, he had been trying to purify himself of his pride (evidently a family defect!). Cacciaguida represented the noble branch of his lineage, the aristocratic figure who allowed Dante to set the tone and confer aristocratic dignity on his family, which, in reality, had very little of it: he had been invested as a knight directly by an emperor, Conrad III of Swabia, and had met his death by following him in the Second Crusade. In the course of that meeting in the heavens of Paradise, Cacciaguida reminds his nephew, who is wandering through the otherworlds, of the Florence of his time, when the city was still enclosed within its first set of walls and people led peaceful, sober and honest lives. He confides to his nephew that there was no ostentation of luxury then, no ambition and no greed for wealth, no corruption of customs that had ruined families. Women were demure and modest then, devoting themselves to their occupations, confident that they would not have to suffer exile or be left alone by their merchant husbands. The cause of the corruption of the "beautiful sheepfold" of Florence, transformed into Satan's evil plant, was due to the "accursed flower," that gold florin, which, having become the standard currency of the time, had fed an immoderate desire for wealth, overwhelming and hardening all minds:

pride and avarice and envy are
the three fierce sparks that set all hearts ablaze.
Inferno, VI

The ladies were crazy about damask costumes with central buttons or enriched with pearl embroidery, or blue wool tunics with flower designs on the front, teal capes and orange hoods.

The greatest amount of fortune came to those who made the most flamboyant clothes, and the orders for those precious garments came from Flanders, England, France and the Kingdom of Sicily, from the royal palace of Naples and that of Spain. More than a third of the 100,000 Florentine inhabitants were employed in the wool business in various capacities: on the highest step were the master wool merchants, that is, the real merchant-entrepreneurs; then came the dyers, the pullers, the *gaulchierai*; and finally, all of the other subordinates, *battilan*i, combers, *divettini, scardassieri*, as well as the large number of people assigned to the collection of semi-finished products and the delivery of the final product.

Hundreds of thousands of florins began to arrive in the city, and a new economy was born, which had little to do with the landed economy of the magnates: it was a river of money that broke free from material goods to become pure finance. With the Peruzzi, the Spini, the Portinari, the Salimbeni and the Adimari, the first banks were established for the purpose of dealing with the robust credit requests of the rich wool merchants and, at the same time, acting as guarantors of all those transactions connected with the sale of the large consignments of wool coming from Castile, Cornwall, the Welsh Marches or the English countryside areas of Gloucestershire or Worcestershire. Among the customers of these banks were popes, cardinals, dukes and marquises, princes of Germany, and the whole court of the king of France.

Thus, the "Arts" were born, real corporations with the aim of protecting the various trades. Not only did they intend to give a sense of belonging to the new productive classes, but also to stand in contrast to the undue pressure of the old oligarchic class. The first of the Arts, needless to say, was that of Calimala, bringing together those who imported raw cloth from Flanders in order to finish and re-export them again. This was considered a major Art, on par with those of Judges and Notaries, Moneychangers and Bankers, of Wool (another one!), of Silk and of Furriers, even, ultimately, that of Physicians and Apothecaries (to which Dante belonged). These were the seven richest Arts, flanked by the fourteen minor Arts, which included all of the common trades, such as masons, bakers, wine makers, butchers, gravediggers, and so on.

At the Roots of Hatred

Marriages in Florence had always been used to heal old wounds and bring together rival families: it cannot be said that this was an infallible tactic, but when it was undermined at its base, damnably tragic consequences could result. This was what happened in 1216, when a private quarrel between two factions turned into a real political conflict, the one that, in simple words, saw the famous "Guelphs" and "Ghibellines" opposing each other for the first time.

The protagonists of the affair were two families, the Buondelmonti and the Amidei, who, in order to put an end to ancient disagreements, had decided to marry the scion Buondelmonte de' Buondelmonti to the young daughter of Lambertuccio Amidei. But love, the real thing, got in the way: Buondelmonte, who had fallen in love with a girl from the Donati family, decided, without warning, to call off the wedding, which had been fixed for February 10. 1215, going that same day to the house of his new fiancée and, adding insult to injury, passing brazenly in front of the church where the bride was waiting for him at the altar.

"Only death can wash away such an insult," cried the Amidei, and, having hastily summoned the consortium to which the Uberti, the Fifanti and the Lamberti also belonged, they decided that revenge should be carried out in a ruthless and public manner. With wise direction, they chose the times, places and symbols of that murder. Everyone in the city knew that something terrible would happen and, while waiting, there was nothing else to do but talk about the marriage gone wrong and Buondelmonte's new love: there were those who sided with the offended family and those who dared to break with the empty custom of arranged marriages, enriching that babel of comments with all of the thousands of linguistic nuances for which Tuscany is still known today. Two great blocs were created: on the side of the Amideis stood most of the ancient noble families, while the Pazzi and Donati families supported the young Buondelmonte and his family.

On Easter Day 1216, Buondelmonte, fully dressed and unaware of the destiny that awaited him, came to the city from his estates beyond the Arno on a superb white palafreno: it was the day of his marriage to the young Donati. He rode through the south gate of the city and continued along the main street towards the church of Santo Stefano, but, near the Ponte Vecchio, hidden behind a statue of Mars, Lambertuccio degli Amidei, Mosca de' Lamberti and Schiatta degli Uberti were waiting to ambush him: the latter, after hurling vicious insults at him, knocked Buondelmonte from his horse with a blow of his club, leaving the other two to stab him to death. Finally, it was Oddo de' Fifanti, also one of the conspirators, who slit his wrists.

"Cosa fatto capo ha," Mosca de' Lamberti had said at the meeting at which they had planned the murder, and that large bloodstain spreading in the street dust was there to certify it. While the assailants were dispersing into the alleys, the relatives of the victim came screaming from the nearby church of Santo Stefano, and Buondelmonte had just enough time to mumble a few words before expiring in his father's arms. The corpse was laid on a wagon near his betrothed and carried around the streets of Florence. That tragic procession was preceded by old Buondelmonti, who, with a hoarse voice, shouted with all the energy in his body at every street corner: "Vendetta, vendetta, vendetta!"

At the news of the murder, general indignation broke out and spirits were inflamed beyond measure, and it was thus that a cursed private story, a rambling affair, soon turned into a real political conflict. In fact, the strong bond of loyalty that bound the Uberti to the emperor meant that the various city alignments that supported the two families involved immediately became linked to the great disputes between the Papacy and the Empire that was inflaming the rest of Italy. Forty-two families of high lineage declared themselves Guelphs, while twenty-four were Ghibellines, and thus was born, in Florence, the war between these two factions that split the fabric of the city forever and bloodied its alleys for several years.

No Holds Barred

From the political point of view, Frederick II, of the noble family of Hohenstaufen, had quickly become perceived as the black beast of the Papacy, for the fear that, with control of southern Italy and Germany already in his hands, he wanted to unify the two domains, including central and northern Italy, thus decreasing the influence of the pope. The real protagonists on the Italian chessboard were, after all, always the pope and the emperor: the former was revered by the Guelph parties, who feared the emperor's expansionist aims, and the latter gave ear to the Ghibelline parties, who opposed the temporal power of the Church.

In Tuscany, the major rival cities of Florence, that is, Pisa and Arezzo, were firmly in Ghibelline hands and, to prevent Florence from falling into those hands, at the strong suggestion of the pope in 1237, a Milanese, Rubaconte da Mandello, was called to the position of podestà. He managed to keep things running smoothly for almost two years, but events began to precipitate in 1239, when Federico II asked Firenze to send troops to support the expansionist campaign he had launched in Northern Italy. The Firenze municipality was divided and, within the Fiorentina militia, violent disorders broke out immediately between Fifanti and Giandonati, the first being in favour of sending troops, the second strongly contrary. The clashes ultimately led to the expulsion of Rubaconte. The most prominent Guelfi then decided to voluntarily leave Firenze, but, when they gathered in the countryside near Siena, they were confronted and routed by the Imperial troops.

From then on, it was a continuous war bulletin, in a mad swing of losers and winners. At first, there was a long list of defeats on the part of the Guelphs. Their discontent began in 1240, with the installation of a new imperial captain in Tuscany and the introduction of taxes intended for imperial pockets. In 1241, the discontent broke out into urban guerrilla warfare and violent armed clashes: the Guelph family of Giandonati attacked the Amidei tower, which housed the pro-imperial captain, but the assault was repulsed. In 1242, it was the turn of the Adimari, also Guelphs, who chose to attack the Buonfanti tower: on that occasion, the bishop interceded, managing to stop the clashes and impose a peace between the two warring parties. In 1248, the Guelphs tried again to turn things to their advantage, putting the city to fire and sword: several Ghibelline towers fell and, with them, those who defended them.

Stupor Mundi

The first half of the thirteenth century was dominated by the multifaceted personality of Frederick II, King of Sicily, Duke of Swabia and Emperor of the Holy Roman Empire, also known as Stupor Mundi. An extraordinarily cultured man, he made Sicily and Southern Italy the centre of a world in which all the great cultures of the time met and blended together: Latin, Greek, Germanic, Arab and Jewish. Moved by an inexhaustible curiosity, he was interested in philosophy, mathematics, natural sciences and medicine; among his favourite pastimes was the hobby par excellence of monarchs and princes, hunting with falcons, on which he wrote a famous manual. To further increase his reputation as a refined sage, he created a long desired Sicilian school of poetry at his court, an initiative later greeted with great enthusiasm by Dante himself. We encounter Frederick II in the sixth circle of Hell, the one reserved for heretics who lie in arks red-hot with fire; he is mentioned en passant by Farinata degli Uberti—a key figure in Florentine politics in those years, whom we will get to know better later—at the end of his speech:

'I lie,' he answered, 'with a thousand, more.
Enclosed beside me is the second Frederick.
Cardinal Octavian, too. Of others, I keep silent.'

Inferno, X

He is mentioned again in the heartbreaking words of Pier delle Vigne, his trusted advisor and minister, who fell into disgrace with the emperor because of the envy of the courtiers and ultimately died by suicide:

I am the one who held in hand both keys
to Federigo's heart. I turned them there,
locking so smoothly and unlocking it
that all men, almost, I stole from his secrets.

Inferno, XIII

However, this success was short-lived, because the Ghibellines hastily asked for help from outside, and the army that arrived in the city a few days later once again routed the Guelph troops, killing their leader, Rustico Marignolli. It was on that occasion that the practice of knocking down the towers of the defeated began: that of the Toringhi on the piazza del mercato vecchio, undermined at its base, collapsed to the ground like an animal struck dead; thirty six other towers met the same fate. The most prominent Guelphs suffered exile, the confiscation of property, the looting of houses and the crumbling of bell towers. Those great mountains of dust and debris, those piles of stone deliberately left in the street were meant to be a severe warning of the high price paid by the vanquished.

Then, the wheel of fortune turned, and it became the Ghibellines' time to drink the bitter cup of defeat. The pro-imperial front had begun to show its first cracks in 1249, when Enzo, son of Frederick II, was defeated at Fossalta and taken prisoner by the Bolognese. To this initial debacle were added other military reverses: in fact, all the expeditions organized by the

Ghibellines to flush out the Guelphs who had taken refuge in the castles scattered throughout the Florentine countryside were accompanied by crushing defeats. In the meantime, out of a pure spirit of opposition, all of the Arts had proclaimed themselves Guelphs, even though they had no particular sympathy for either the pope or his political aspirations, but only for the generosity of his purse.

To try to put an end to the continuous civil war that poisoned the air every day, it was decided that the governing bodies would be changed: thirty-six citizens, six per district, with sympathies for neither Guelph nor Ghibelline, with the consent of the entire population, met in the rooms of the towers of Marignolli and Anchioni, near San Lorenzo, to write the new city decalogue. On October 20, 1250, the political order of the so-called Primo Popolo was promulgated: alongside the old office of podestà, the new figure of Captain of the People was instituted, a man who would be given command of the city's army. The not-so-subtle intention was to bring the actions of the podestà, often too close to the interests of the Ghibelline noble families, under popular control. This double magistracy was assisted by a council of twelve "elders" and by other governing bodies occupied by those in the city who had the most money: the aim was to supplant the old aristocratic oligarchy with a new and rampant oligarchy of census.

Things then changed definitively in December 1250. "The Sun of the Universe that shone in the midst of the people has fallen": thus did Manfred communicate to his half-brother Conrad the death of their father Frederick II. At the news, Pope Gregory IX gave the order to ring the bells of all Christian churches, to announce the death of the antichrist. It was December 13, 1250, the feast of St. Lucy. Frederick II had died in Torremaggiore, in Apulia, under the exact circumstances that his master of esotericism and magic arts, Michele Scoto, had predicted: near an iron door and in a place located sub flore. This prediction had been enough to cause Frederick to deliberately avoid going to Florence (Florentia) for years and yet perhaps fate was destined to have its way, because, taken ill during a hunting trip in the Apulian countryside, he was rushed to a castle and laid on a bed behind a tapestry. At the insistent requests of Frederick, who had strange premonitions, his men proceeded to remove the tapestry, discovering, with great surprise, that it hid an iron door upon which was engraved a flower... It is said that, while dying, the emperor whispered "Post mortem nihil est," after death nothingness, as if to seal the reputation of unbeliever, materialist and epicurean that the pope had sewn on him.

The disappearance of Frederick II left all of the Ghibellines of Italy orphaned: taking advantage of that particular moment of disbandment, in Florence, the Guelphs once again put their hands to arms and finally managed to get rid of the Ghibelline nobles, banishing them from the city and ruthlessly pulling down the towers of their palaces. The fugitives found refuge in the castles of Montevarchi and Romena, in the countryside near the Ghibelline city par excellence, Arezzo.

Florence also opened hostilities against the major pro-imperial centres of the region, among which were Pisa, Pistoia and Siena. Folded in 1253, Pistoia was obliged to recall the Guelph exiles; in 1255, Siena had to accept the humiliation of a treaty that prevented it from giving asylum to the partisans of the Empire.

Ten years of relative tranquillity then passed in Florence, broken in 1258 by Manfredi, who resumed the warlike operations in an effort to complete the expansionist aims of his father, Frederick II. The Ghibellines who were scattered throughout the peninsula rejoiced: in Florence, they promptly organised a conspiracy to overthrow the government. At the head of it were the Uberti, one of the few Ghibelline families still present in the city, having escaped the ban of eight years earlier. The conspiracy was discovered, however, and the Guelph reaction was prompt and ferocious: Schiatta degli Uberti died with his weapons in his hand, pierced by arrows; another Uberti had his throat slit in the market square; an Infangati had his head cut off; the bishop of Vallombrosa, Tesauro Beccaria, was assassinated for suspected anti-Guelph activity. Those who managed to escape the massacre, led by Farinata degli Uberti, took refuge in nearby Siena. All of the conspirators had their houses burnt, their roofs blown off, their furniture destroyed, their safes plundered and their horses stolen: Florence was once again studded with corpses and rubble.

Farinata Degli Uberti

Manente degli Uberti, known as Farinata because of his platinum blond hair, is one of the most vivid figures of Dante's *Inferno*, one to whom Dante paid great homage with a memorable description of his uncompromising political commitment even amidst the infernal flames and the imposing and proud demeanour with which he nurtured a superb contempt for the world of the damned around him. Belonging to one of the oldest and most important Florentine families, Farinata was one of the historic leaders of the Ghibelline side of Florence. He took refuge in Siena after the expulsion by the Guelphs from Florence in 1251, and became the main protagonist of the crushing Ghibelline victory of Montaperti, a battle that took place in the Sienese countryside on September 4, 1260, during which more than 10,000 Guelph soldiers died despite fighting against only 600 Ghibelline soldiers. The battle was characterized by the betrayal of the Florentine Bocca degli Abati, who, although he was on the side of the Guelphs, was actually a Ghibelline fifth columnist: when he saw the Ghibelline troops arrive, Bocca degli Abati did not hesitate to cut off the hand of the Florentine standard bearer, causing chaos and bewilderment in the Guelph troops, who were then left without a guide to tell them what to do and where to go. With that, the charge of the German cavalry on the Florentine troops began:

The massacre, the mindless waste
that stained the flowing Arbia with blood

Inferno, X

At nightfall, the Ghibelline commanders gave orders to save the life of those who surrendered… so long as they were not Florentines, who were to be killed upon capture under any circumstances. The flags and banners of the Florentines were taken, and the banner of Florence itself was attached to the tail of a donkey and dragged through the dust. At the Diet of Empoli, which followed the battle of Montaperti, Farinata nobly demonstrated his great love of country by standing up to the deputies of Pisa and Siena, who wanted to raze the city of Florence to the ground. However, this ultimately failed to rehabilitate him in the eyes of the Florentines. He died in 1264, which opened up his heirs to having all of their property confiscated, but the revenge would not spare even the dead: the remains of Farinata and his wife Adaleta were exhumed in 1283 to answer the charge of heresy and burned once the verdict of guilt was reached. The episode left a very strong impression on Dante, then just eighteen years old, who mentioned it with admiration in the *Inferno*.

International Intrigue

"Has the donkey of Porta brayed?" cried Corso Donati, as soon as he awoke, to those who stood under the window of his palace, alluding to his great enemy Vieri dei Cerchi. For some time then, the Ghibellines had been definitively eradicated from Florence, the battle of Campaldino, fought and won by the Guelphs in Poppi's campaign of June 1289, having sanctioned the new political course, but one cannot say in all honesty that peace had returned to the city: the struggle had simply shifted to another front.

The Guelphs were, in fact, divided into two major parties, one headed by Corso Donati, the other by Vieri dei Cerchi. Feeding the fight without quarter between these two partisanships was Pope Boniface VIII: not accustomed to divine practices, but very attentive to earthly ones, for several years, he had been hatching the idea of extending his reach into Florence, to annex it into the pontifical possessions, or at least to subject it to his will through Charles of Valois, brother of Philip the Fair, King of France. What appeared to be a violent dispute in the city was, in reality, the visible tip of a great international intrigue, an affair in which Dante Alighieri also took part, and which, in the end, caused him to be crushed. It is worth taking a few steps back to outline the plot of this very complicated game played out on the Florentine chessboard.

Although formally banned from Florentine political life, in reality, the original magnates, those with a castle in the countryside and a turreted palace in the city, had undergone a singular social transformation: over the years, they had formed relations with the bankers, the wool industrialists and those who had large estates in the territory, and now enjoyed positions of great economic and financial privilege, using them for their own manipulations. Thus, a violent political struggle arose between the fat people, who included the magnates mixed with the rich bourgeoisie, and the petty people, those of the major and minor Arts.

An authoritative witness of those turbulent times, Dino Compagni, wrote that:

The aristocrats and the great citizens, who were unsatisfied, used to do many insults to the common people, beating them and using other insults. So many good citizens, commoners and merchants, among whom was a great and powerful citizen (a wise, brave and good man, called Giano Della Bella; very lively and of good stock, who was sorry for these insults), who made himself their leader and guide [...] they encouraged the people.

It was therefore Giano Della Bella (a.k.a. Janus), a member of one of Florence's oldest and noblest Ghibelline families who had become a Guelph and later a champion of the city's popular classes, who led the revolt against the magnates in 1292. He had the right physique du rôle: he was courageous, energetic, full of charm, irresistibly eloquent and, above all, angry.

He became prior, and succeeded in pushing through an epoch-making reform, the so-called Ordinamenti di Giustizia (Orders of Justice), promulgated on January 18, 1293, establishing that, in order to be eligible for political office, enrolment in an Arte was necessary; above all, it was decreed that, without exception, magnates would be excluded once and for all from governing the city. No distinction was made between the Guelph and

Ghibelline aristocracies: first of all, the terms regarding who was to be considered a magnate were written in black and white, but then the names of the 73 dynasties in the city and 74 in the hinterland were explicitly given, robbing them of the possibility of electing one of their members to any of the offices of the Signoria. On this list were the most famous names in Florence: the Uberti, the Donati, the Frescobaldi, the Cavalcanti, the Buondelmonti, the Bardi, the Lamberti, and many others with whom these families had relations, including those of the upper middle class. For those newly rich who had resisted the temptation to ennoble their lineage with that of the aristocracy, the various Peruzzi, Medici, Pitti and Strozzi families, the Ordinamenti were, instead, the beginning of their fortunes.

Thanks to the Ordinances of Justice, the nobles lost the arrogance with which they had ruled in previous years, when they would go out into the alleys with their squads and beat anyone within their reach with impunity: now, they could be denounced by any commoner, arrested and thrown into the prisons near the Bargello. "From now on, beware of the tails of your horses; if one of them touches the face of a passer-by, you are lost," said a magnate. The people participated with passion in the demolition of the houses of the magnates, in the spoliation of their property and in enjoying their ruin.

The reaction of the magnates to that unexpected course of events was unbalanced at first, and paradoxically resulted in a series of murders within their own consortium, a reckoning that reddened the alleys of noble blood: two Amidari were slaughtered by the Della Tosa, a Gherardini ended up stabbed by the Manieri, the Velluti avenged an old wrong suffered by killing some Manelli. Having satisfied their thirst for revenge, the finest heads of the magnates did, however, understand that there was only one solution to their problems: get Giano Della Bella out of the way, by hook or by crook. It was not an easy problem, because Giano enjoyed tremendous support among the people: he had ordered the paving of the Ponte Vecchio and the reclamation of the unhealthy wells, had created Cascine Park, had the Baptistery covered with the white and green marble of Prato and, unusually, he did not steal and did not let anyone else steal.

Their first idea was to kill Janus, but they preferred to fall back on another solution, for fear of popular anger. So, they opted for a vast operation of discredit that would rob Janus of the trust he enjoyed, ignominiously exploiting certain exaggerations of some of his missteps. The worst rumours were immediately spread about him, and the defection of his closest collaborators, Lippo Velluti and Caruccio del Verre, was brought about to the sound of falling florins, until the key episode occurred that made the discontent with Janus grow beyond measure until he was expelled from Florence.

The protagonist of this episode was Corso Donati, who made his public debut in a street brawl in which he wounded his cousin and killed the latter's servant. However, despite the many witnesses present at the murder, no one dared to breathe a word when, in front of the potestà, Corso Donati shamelessly claimed that it was his cousin who had killed his servant so that the blame would fall on him. While that hearing was in progress, another Donati, named Novello, killed a commoner in the street and, upon being arrested, was immediately sentenced to death. But the best was yet to come: Corso Donati appeared in court, surrounded by his masnadieri, eventually exiting with Novello, freed upon a verdict of full acquittal signed by the potestà!

Immediately, furious riots broke out: in his rush to try to put an end to that violence, Janus was instead overwhelmed, accused by the people of engaging in embezzlement with someone like Donati. Janus's luck with the people was now at its sunset: his enemies had finally obtained their goal. At this point, he preferred to flee abroad, pursued by a death sentence resulting from a trial in absentia, a verdict that was quickly reversed by his former friend Lippo Velluti, who had become prior.

Having achieved their aim of ousting Janus, the magnates attempted a coup d'état, but were stopped at the last moment by the protests of the people; they were subsequently content to settle for an amendment to the Orders of Justice, which at least partly watered down the original spirit. It was at this point, in the last decade of the thirteenth century, that the neverending feud began between the Black Guelphs, led by Corso Donati, and the White Guelphs, under the command of Vieri dei Cerchi. The Cerchi had created an immense fortune through trade, while the Donati had less wealth, but an enormous pride of lineage. All the nobles of blood gathered around the Donati, while the Cerchi had the support of the new bourgeois classes that had flourished in Florence with trade. It was therefore inevitable that the fierce rivalry between these two families would immediately take on a political colour.

Dante Alighieri, who, in those very years, entered the agonies of the city's public life, found himself involved in that intriguing conflict without quarter that was destined to stain the streets of Florence with blood for years.

Corso Donati, called the Baron, was unscrupulous and cynical beyond all limits, pugnacious and overbearing, proud and thirsty for power: he was the brother of Forese Donati, with whom Dante had the famous, previously mentioned poetic quarrel, as well as the cousin of Dante's wife, Gemma Donati. However, in spite of these ties of kinship, Dante's sympathies went to the White Guelphs, because he could not bring himself to tolerate that delinquent Corso Donati and his arrogance. Stories about him abounded: it was said that he had poisoned his wife (one of the Cerchi) after she had gratified him with a rich dowry; it was known that he was directly responsible for many unsolved murders in the city, so much so that he had the nickname of "Malefami," just as it was known that he had not hesitated to kidnap his sister Piccarda from the convent where she was living as a nun to give her in marriage to Rosso Della Tosa for bizarre political interests. Moreover, he was famous for his outbursts of anger when someone disobeyed him and he never hesitated to get physical in order to impose himself. He had been the absolute master of the district of Porta San Piero for years, but his fame had been obscured when the Cerchi family arrived and transformed the palace they had acquired in that quarter into a sumptuous place, frequented by men and women of high society, who were welcomed by servants in livery. The hatred between the two neighbouring families had resulted in ambushes, mutilations, assassination attempts, and killings on both sides, all of which went on for several years.

In the year 1300, Corso Donati, to escape a penal procedure, went to occupy the role of potestà in nearby Pistoia, thereby momentarily leaving the field free for a temporary redemption of the Cerchi. The timing, however, was not the best: in Florence, it was rumoured that, as mentioned earlier, Pope Boniface VIII was trying to reach an agreement with the King of France, Philip the Fair, to conquer the city thanks to an armed intervention by his brother Charles of Valois. To address this danger, the Cerchi adopted a contradictory and temporizing policy that ended up bringing them to ruin: on the one hand, they publicly opposed papal interference in Florence, but, on the other, they, in fact, sent Florentine troops to help the pope in the wars against his internal enemies, such as the Colonna (with their city of residence, Palestrina, being razed to the ground) and the Aldobrandeschi in Maremma. In May 1300, a delegation was sent by the Cerchi to Rome, officially for a diplomatic visit, but with the secret task of identifying the names of those Florentines who supported the pope's policy and were passing along information to him about what was happening in the halls of the priori and other organs of the Signoria. Three of the four Florentines identified and accused of treason had their tongues cut out, despite the vehement protests of Boniface VIII. In light of these last negative developments, the pope hastened to set up his countermeasures, approaching Corso Donati and making him his ally: the violent nature of the character could, in fact, be useful to him.

At the beginning of May 1300, a furious brawl broke out following an act of aggression by a group of Cerchi against the Donati, who, a week later, tried to seize power with a coup d'état. However, the attempt failed, and Corso Donati, considered to be the instigator of the sedition, although, at the time of the events, he was in Rome with the pope, was condemned to death and his house partially destroyed. The situation remained explosive, and Boniface VIII offered to act as peacemaker, although his true motive was insidious: he sent one of his cardinals to Florence whose real task was to further muddy the waters.

It was in that fiery summer of 1300 that Dante was elected one of the six priors of the city, having previously joined the Arte dei Medici e degli Speziali. This was the origin of all his subsequent misfortunes, as he later wrote. He had not yet had time to take office when, on June 23, 1300, on the eve of the feast of San Giovanni, more disorder broke out in the city. In an attempt to re-establish calm, the priors decided to expel the worst troublemakers, both those of the white Guelphs and those of the blacks. The blacks were confined in an Umbrian castle, while the whites were forced to go to Lunigiana, a swampy region near Sarzana. Among the latter was Dante's dear friend from his youth, Guido Cavalcanti, who fell ill with malaria and died in August 1300, a memory that haunted Dante even into the years when he wrote the *Inferno*, in which he meets the melancholy Cavalcante de' Cavalcanti, the Stilnovist poet's father.

Dante's real misfortune, however, was to find himself in the path of the political-territorial machinations of Boniface VIII, the most fanatical proponent of the temporal power of the Church and of the idea that everyone should be subject to the Roman pontiff. He was also the greediest of men, as was seen when, in 1300, he offered a "full pardon" to those who went to Rome in the sacred year of the Jubilee, proclaimed for the first time in the history of all Christendom. An extraordinary crowd of penitents subsequently rushed into the Holy City, carrying a mass of money so overflowing that the sacristans were forced to gather it up with a rake. Among those who visited Rome during the Jubilee was Dante, who, dismayed at the chaotic and whirling spectacle, watched the slow pace of the streams of pilgrims and heard a Babel of languages resounding in his eardrums, perhaps drawing inspiration for his journey to the otherworlds.

The tension between Florence and the pope continued to grow day by day, as did the mutual distrust: the white Guelphs, then in power, were suspicious of a pontiff who was so interested in earthly goods, particularly those of

Florence; for his part, the pope saw their explicit hostility, which convinced him that consolidating his relationship with the blacks was a good idea.

> **Devil of a Pope**
>
> Dante's hatred for Pope Boniface VIII was so intense that he predicted his eternal punishment, to be administered upside down, in the *Inferno*: Benedetto Caetani, who had become pope in 1294 following the clamorous abdication of Celestine V (perhaps due to the mysterious intrigues of Boniface himself), was, by right, among his worst enemies, since Caetani was considered responsible for the corruption of the Church, the fall of the White Guelphs in Florence and, therefore, Dante's exile. Had he not been deceitfully detained in Rome while Corso Donati and his scoundrels were burning the city to the ground in Florence? The mere mention of Caetani's name was enough to make Dante go into a frenzy: the Roman curia of the time of Boniface VIII was to be categorized in *Purgatory* as a "loose whore" while, in *Paradise*, Saint Peter would allude to him with rough words:
>
> *He who on earth has robbed me of my place,*
> *[...].*
> *of my own burial ground a shit hole*
> *reeking of blood and pus.*
>
> Paradiso, XXVII
>
> Boniface VIII affirmed, in the most energetic and intransigent terms, the theocratic ideal according to which the spiritual power exercised by the pope had the task of instituting temporal power and judging it, while the work of the pontiff could only be judged by God. It followed that anyone who opposed spiritual power was, in fact, opposing God's will. Combined with the worldliest desires for wealth, these political principles made Boniface VIII one of the most greedy, proud and cruel figures in the history of the Church. The ruthless fight against the Colonna family, with the total destruction of the town of Palestrina, their stronghold; the unscrupulous manipulations and the sinister interference for control of Florence: all of this testifies to his brutal conviction of being the supreme authority on earth. He was the last great medieval pontiff, destined, however, to succumb to those who were even more criminal than he: excommunicated by Boniface VIII in 1302, the king of France, Philip the Fair, reacted by accusing him of heresy, sodomy and demonic practices, ordered his emissary William of Nogaret, as the head of an army, to march to the papal residence in Anagni. Boniface VIII was humiliated and imprisoned. Later freed by a popular uprising, he could not withstand the outrage: he died after only a month, still foaming with rage.

The pontiff's next move was ingeniously disruptive, a move that definitively upset the cards on the table, creating an avalanche that would end up overwhelming the White Guelphs forever: he appointed Charles of Valois as his peace broker on Florentine affairs and invited him to enter Florence with one of his armies. Corso Donati, on behalf of the Black Guelphs, rushed to visit the Pope's emissary in Siena on October 16, 1301, offering him all his support, plus a large sum of money to fund that "mission of peace." For their part, the Whites hurriedly sent three ambassadors to Rome to ask the pope to recall his emissary or, at worst, to at least gain assurances that Charles of Valois would respect the Florentine constitution. Within that delegation, saddled with such an impossible mission, was Dante Alighieri. His sense of despondency is summed up in the questions he uttered for the occasion: "If I go, who remains? If I stay, who goes?".

The pope coldly welcomed the trio, or, rather, told them that it was better that the rulers of Florence submit to his authority, then quickly dismissed two of them, keeping the third, Dante Alighieri, in Rome, his spies having told him that Dante was one of his most implacable enemies. In the meantime, Charles of Valois had arrived at the gates of Florence, subsequently entering it on November 1, 1301, with great pomp, accompanied by two hundred Tuscan knights, who had come to honor the emissary of the pope, brother of the king of France. The priors had no choice at that point but to surrender control of Florence to him: Charles of Valois solemnly swore that he would maintain peace.

But the promise lasted only three days: on the third day, Corso Donati arrived in the city with an army of black Guelphs who had previously been exiled, armed to the teeth. They passed through the gates garrisoned by the French and headed for the prisons, where they freed the common criminals, welcoming them among their troops. After that, they started looting and initiated an orgy of unprecedented violence that lasted for six days, during which Florence was put to fire and sword.

The next step was to annul the sentences that hung over the heads of the blacks, and to set up courts quickly for the purpose of condemning their bitter enemies. Many Cerchi, even those with the least amount of suspicion hanging over them, in an effort to save themselves, betrayed their own, going over to the side of the black Guelphs. Hundreds of death sentences were pronounced in the blink of an eye, with those who managed to escape the executioner preferring to flee, in an exodus of biblical proportions.

The purge of the white Guelphs continued unabated, and the judicial campaign also ultimately affected Dante, who was still being held in Rome by Boniface VIII: accused of embezzlement in public acts and initially condemned in absentia to a very high fine, given his continued absence from Florence, the judges had no scruples about toughening his sentence to death by burning. His house was stormed and demolished. Thus began his sad life as an exile.

How Salty Other People's Bread Tastes

Dante never returned to Florence. With the end of his political career, a wound was opened in him that was destined never to heal. He also had to move about with extreme caution, because, with a death sentence on his shoulders, he could be killed, legitimately and with impunity, at any time by anyone with ill intentions towards him. He left everything behind, family, possessions, friends, and began to wander from patron to patron, from court to court, always in search of a benefactor, a reliable protector, a powerful person who was also his friend.

In the early years of his exile, he came into contact with other White Guelphs who had taken refuge in Arezzo, and, with them, he tried to organise a military campaign to defeat the Black Guelphs and return to his homeland. For this purpose, he went to seek help in Bologna, and then in Verona, but it was all in vain: the last fatal sortie to Lastra a Signa, on the outskirts of Florence, made in July 1304 by a ragtag army of white Guelphs, was easily routed and the prisoners were paraded through the streets of the city on the back of a donkey before being decapitated by the executioner. Dante abandoned that "wicked and foolish company" as well inconclusive as the White Guelphs who had escaped and from then on he did "part for himself", namely he became a solitary soul.

The events in which he was involved after 1304 are only partly known, with many details, in fact, being masked by the legends that began to arise about him. Certainly, his finances were in a disastrous state, and thus to the gloomy solitude was added the most horrible poverty. In a letter written in November 1304 to the Counts Oberto and Guido da Romena, he let himself succumb to notes of deep despair:

> *Poverty, like a vindictive fury, has driven me, deprived of horses and arms, into her prison den, where she has set herself incessantly to hold me hostage; and though she struggles with all my might to free me, she has so far got the better of me.*

And, more or less in the same period, in the pages of the *Convivio*, all of the bitterness that lacerated his soul during his wandering through Italy came out:

When it was the pleasure of the citizens of the beautiful and famous daughter of Rome, Florence, to throw me out of her sweet bosom—in which I was born and nourished to the very summit of my life, and in which, with good grace, I wish with all my heart to rest my weary soul and finish the time that is given to me—I went wandering, almost begging, through almost all the parts to which this language extends, showing against my will the scourge of fortune, that is often unjustly imputed to the wounded.

Truly I have been a ship without sail or steering, carried to various ports and mouths, and tossed about by the dry wind that steams my sorrowful poverty: and I have appeared in the eyes of many who might have imagined me in some other form, perhaps because of some fame: in the sight of whom I not only harm my person, but every work, whether already done or to be done, is of lesser merit.

He spent a couple of years in Bologna, between the end of July 1304 and the beginning of 1306, a period during which he probably wrote the *Convivio* and the *De vulgari eloquentia*, both of which, however, remained unfinished and were never published. Writing, as well as reading philosophy, provided partial consolation for the anger and anguish that devoured him. However, the city of Bologna also rebelled against the White Guelphs who ruled it, and Dante had to flee again, finding refuge in the mountains of Lunigiana, as a guest of the Malaspina, a family later praised in *Purgatory*.

In that period, having nothing more to lose, he humbly tried to convince Florence to revoke his exile, with a gesture of public repentance, a request for forgiveness: he wrote a heartfelt letter that began with, "My people, what have I done to you?", but, unfortunately, it had no effect. He then moved to Casentino, as a guest of the Counts Guidi, and there, to the pains of exile, were added the pains of love: he fell prey, in fact, to a violent passion for a woman from Casentino, blonde, curly-haired, with her head surrounded by garlands, immortalized in the verses of a song called "Montanina," which he sent to his patron, Moroello Malaspina. He paid great attention to the development of other military campaigns set up by the White Guelphs to try to overturn the political fortunes in Florence, but, one after the other, these foundered miserably. Legend has it that Dante, deeply disappointed, took the road to Paris.

Both Boccaccio and Giovanni Villani, a fourteenth-century Florentine historian, speak of Dante's stay in Paris. Some references suggest that Dante did indeed visit Paris and that he may have attended lessons in philosophy and theology at the Sorbonne, as seems to be suggested by the mention in the *Comedy* of the "Vico de li Strami," or rue du Fouarre, where the Faculty of Arts was located.

In 1308, he was back in Tuscany, and he learned there of the ruinous end of his arch-enemy, Corso Donati: ever more eager for power, Corso had gone too far in his arrogance, thus antagonizing a series of equally violent families, such as the Della Tosa, the Pazzi and the Brunelleschi, who had decided to put a definitive end to his harassment. Arrested, Corso Donati managed to escape from prison and, on horseback, tried to leave Florence as soon as possible. Caught in the district of Rovezzano by the armigers of the Della Tosa family, he fell ruinously from the saddle, but remained bridled in the stirrups. To his embarrassment, the animal dragged him to the ground, where he was then finished off with a lance. With a certain satisfaction, Dante recalls this episode in the *Purgatory*, through the victim's brother Forese Donati:

That beast rampages faster with each step,
accelerating always, till it strikes
and leaves his broken corpse a mangled mess.

Purgatorio, XXIV

The Remains of Dante

As in life, Dante found no peace even when he was dead. His body was initially buried in a marble urn in the Basilica of San Francesco in Ravenna, where his funeral was held. In 1483, the tomb of the poet was decorated with a monument, which, however, fell into disrepair and remained in a disastrous condition for about two centuries, when the interest of the city of Florence in retrieving the bones of its illustrious fellow-citizen became apparent. The people of Ravenna unsurprisingly objected, believing that the Florentines did not deserve the remains of the man whom they had hunted and humiliated. The pressure to accede became stronger and stronger, including due to the direct intervention of the Medici popes Leo X and Clement VII.

When the tomb was opened, the Florentines who had come to Ravenna to bring home the bones of the Poet were stunned to find the tomb empty: the remains had been removed by the Franciscan friars and placed in a wooden box, the location of which remained a secret handed down from generation to generation until 1810, when they decided to hide the box in a walled door in an oratory. And there it remained until 1865, when it was found by a bricklayer engaged in the restoration work of the convent: inside were an almost intact skeleton and a letter, signed by Father Antonio Santi, attesting

> that the box contained the bones of Dante. Once the body was reassembled, it was exhibited for some months in a glass case, and then finally buried inside the small temple built by Camillo Morigia, in a wooden box protected with lead. Since then, apart from a momentary transfer during the Second World War to avoid possible destruction, Dante's bones have never left Ravenna. The Florentines still wait in vain for the return of the man they condemned to go exiled into the world, "like fish to the sea."

But the real novelty of those years was the Italian campaign of Arrigo VII of Luxemburg, who, called directly by the pope, went to Italy to fight as he made his way to Rome to be crowned Emperor of the Holy Roman Empire, thus taking possession of his Italian lands and putting an end to all of the infighting that periodically broke out in the various Communes of the peninsula. Dante had long since come to the conclusion that only a force outside Italian politics could bring calm, and the arrival on the scene of Arrigo VII therefore revived his hopes of ending both his personal predicament and the general Italian crisis. He therefore wrote heartfelt public letters to the various princes of the cities and to his fellow Florentines, exhorting them to welcome the German emperor as the one who would re-establish order in earthly affairs, avoiding, however, any mention of the sophisticated theses expressed in his *Monarchia* (composed in those very years) regarding the dualism that set the Church in opposition to the Empire.

Some Ghibelline cities, such as Milan and Verona, welcomed the emperor with all honours, others, Florence in particular, were instead fiercely entrenched in their opposition to that foreign monarch, fearing the possibility of losing their autonomy and wealth. The talk of Guelphs and Ghibellines began again, even if, this time, it was the Pope himself who had called the German emperor to Italy. To complicate things further, Arrigo VII, more a romantic character than a warmonger, was not very resolute; in fact, he allowed constant delays, advancing, but only in small steps, never giving the impression of an imperial power marching to the conquest of Italy. The emperor was dealing with the siege of Florence when he learned of the pope's sudden change of sides, the latter having decided to abandon the former to his fate, allying with Robert of Anjou, king of Naples, and therefore with the Black Guelphs present in the city. The siege of the city of Florence therefore resulted in a complete fiasco for the entire imperial cause.

The following year, the emperor died of malaria at Buonconvento, while Dante meanwhile repaired to the court of Cangrande della Scala in Verona. With the death of Arrigo VII, Dante's last hope of returning to Florence had also disappeared. In *De vulgari eloquentia*, he wrote: "Nos autem, cui mundus est patria velut piscibus equor," I have the world for a home, as the fish have the sea.

And it was precisely in Verona, a city that he had already had the opportunity to appreciate in the past and, above all, to benefit from the reading of the many ancient volumes that were present in its very rich library, that Dante Alighieri may have crossed paths with Marco Polo: the former came as an exile, the latter arrived as part of some delegation on behalf of the Serenissima. On the other hand, there was room for everyone in that great court, and Cangrande, with his convivial spirit and dedication to the pleasures of the table, surrounded himself with musicians, historians, poets and jesters to cheer up the evenings and entertain his guests, who would discuss politics, art or military battles. The prince seemed to be ahead of his time; in fact, he had all the qualities of a Renaissance prince: if, on the one hand, he was a shrewd diplomat, on the other, he was an intrepid leader, an exquisite host who loved the arts and the pleasures of life.

There has been much discussion over the centuries about the true relationship between Dante and Cangrande of Verona, a curiosity that has flourished, above all, because of the obvious contradiction between the cross-section offered by the many anecdotes that have arisen around the Poet's stay in that city and the lofty tones of Dante's famous epistle to Cangrande, in which he dedicated the last Canticle of his *Comedy*, *Paradise*, to the Lord of Verona.

The anecdotes, particularly tasty because they were based on the art of words of which Dante was an unsurpassed master, were collected and divulged by Petrarch. In one of them, it is told that Cangrande had asked Dante how it was that the court entertainer, although foolish and clumsy, managed to entertain all the guests, while he, who claimed to be a great poet, could not entertain anyone; the reply was not long in coming: "You would not be surprised if you knew that the cause of friendship is the equality of customs and the similarity of spirit."

Another biting anecdote concerned the mortification that Cangrande had tried to inflict upon Dante during a meal, ordering his servants to secretly place at Dante's feet the bones of the game eaten by the rest of the guests. When, as was the custom in those days, the table was moved to clear the room, everyone began to laugh at that mountain of remains under the Poet's chair, the latter, however, being quick to comment: "No wonder the dogs ate their bones. I, who am not a dog, have left mine," with obvious, sarcastic reference to the gentleman's name[1] of Verona.

---

[1] In Italian, Cangrande literally means "Big Dog".

In spite of these outrageous mortifications, there is the famous Epistle XIII to Cangrande della Scala, in which Dante not only dedicated *Paradise* to his protector, but did so with resounding words of admiration and friendship. This first part of the epistle, as is well known, was followed by a second part, in which the Poet set out in some detail the subject of the entire *Comedy*, which is open to both literal and allegorical interpretation. Many scholars have always doubted the authenticity of this missive, which they say is too superficial and schematic compared to the depth Dante displayed as a poet and writer.

It is difficult to establish whether the epistle is false or not, especially in the absence of an original copy, but it is certain that, after several years spent in Verona, perhaps only in the unrewarding role of entertainer, Dante decided to settle in Ravenna, as a guest of Guido Novello da Polenta. According to Boccaccio, this was in the year 1314. As established by Florentine law, sons followed their fathers into exile once they turned fourteen and, in fact, Jacopo and Pietro Alighieri soon joined Dante in Ravenna. They were accompanied by their sister Antonia, who later became a nun in a convent in Ravenna, taking—as chance would have it—the name of Sister Beatrice. There is no certain information about his wife Gemma, but most people think that she probably stayed in Florence and never rejoined the rest of the family.

In Ravenna, as well as working hard on the poem of life, he also had the opportunity to carry out diplomatic missions on behalf of Guido Novello. The last of these, in 1321, involved his traveling to Venice to negotiate a rather complicated and delicate matter concerning the scuffles that had taken place between some sailors of the two cities over salt pans near the mouth of the Po. It seems that, in Venice, fearing the strength of his eloquence and his persuasive abilities, the high officials of the Republic prevented the Poet from speaking to the Doge.

All that was left for Dante was to return to Ravenna, first on rafts, through the Comacchio lagoons, and then on a cart, through the vast river marshes. It was in those unhealthy areas that he contracted malaria, just as had happened to his friend Guido Cavalcanti and his emperor Arrigo VII. He arrived home in a feverish state, welcomed by his children. He died on the night between September 13 and 14, 1321, no doubt with the image of the Rose of the Blessed of Paradise before his eyes.

Astrolabe and Compass

In Which We Learn of Noble Castles, of Caravanserais, of Numbers and Giants, of the Stars of the South and of Women's Smiles

It All Began in Jerusalem

There was never a more compelling or more complete journey through the knowledge of humanity and things than those of Dante and Marco, the former in the labyrinths of the mind, the latter along the paths of the world.

Dante travelled from Jerusalem to the centre of the Earth, climbed the mountain of Purgatory, located at the antipodes, and finally, in the ten heavens of the Aristotelian and Christian cosmos, as far as Empyrean, reached the sapientia concentrated in God. His was a cosmic journey that looked at the natural and human worlds through a lends of total poetic osmosis, enclosed in a work that was born of precision, but that drew, above all—think of Leopardi—from the margin of vagueness and imagination that poetry placed at the disposal of the poet and his imagination. And it is in this space between vagueness and precision that our hunger for knowledge is measured, a hunger that Dante satiates in the *Comedy*. It is not, of course, the same knowledge as that of the merchant Marco, a navigated man of the world, who, in his twenty-seven years at the court of the Great Khan, travelling and observing the nature of men, "knew more of those things than any man that was born."

In the pages that follow, we will accompany our two wayfarers more closely during their journeys and, in giving a fresco of the fabulous period that was the fourteenth century, we will try to ideally synchronize these two extraordinary journeys of theirs. We will therefore oscillate from one to the other—from the otherworldly and ethereal journey of the Argonaut Dante Alighieri to the earthly and dusty journey of the explorer Marco Polo—stopping to reflect on the vast archipelago of themes that accompany such singular journeys: we will speak of the noble castle of the IVth Canto of the *Inferno* as of the minarets of Baghdad; of the monsters that populate the infernal caverns as of the dragons that crowd the Chinese imagination; of the suggestive female figures of the *Comedy*, but also of the seductive women of the Asian villages; of the astronomy of the cosmos and the geography of the Earth; of the mad flight of Ulysses, but also of the slow, long caravans along the Silk Road; of the astrolabe and the compass, indispensable instruments of orientation and navigation in an age that relied on the sun and the stars to know the time and the routes to take.

First, however, it is worth remembering the singular kinematics of these journeys, to appreciate the vastness of the times and spaces involved. There are obviously some uncertainties about the dates, real or imaginary, of the various stages of these journeys, but we will gloss over these, simply taking the most plausible ones for granted.

Dante's otherworldly journey, as we have seen, begins on the night of March 25, 1300, and ends on the night of March 31 of that same year, a week that coincides with the Easter period. 1300 is also the year of the first Jubilee, proclaimed by Pope Boniface VIII, providing remission of the penalty for sins committed to those who go on pilgrimage to Rome.

Having entered the infernal chasm that opens up at the foot of Jerusalem, Dante, together with Virgil, takes three days (the same number as for Christ's resurrection) to cross the entirety of Hell, arrive at the centre of the Earth and then exit into the suffused sky of dawn on the beach of Purgatory, crossing the "natural burella." Even if we take into account the various ferries, including the one over the Malebolge precipice effected on the back of Geryon, we are talking about a distance of 12,700 km (the diameter of the Earth) covered in a space of thirty hours or so, on foot! Obviously, it seems like a joke, unless we invert the reasoning and take for granted the duration of the journey into the bowels of the Earth: this allows us to deduce what the estimated length of the radius of our planet was at Dante's time. By doing so, the calculation of the Earth's radius gives us, roughly, an estimate on the order of a few hundred kilometers, which is also the height of the mountain of Purgatory, since Dante takes care to emphasize that the origin of the mount is due to the retreat of the earth when Lucifer was cast out of Paradise.

It should be noted that Dante and Virgil take the same three days to go from the base of Purgatory to its summit, where they find the Earthly Paradise, although this distance is only half of the journey made by the two poets in crossing the Earth from side to side: the reason for this is not only that the journey along the mountain is uphill, but also that, while climbing along the crags of Purgatory, they decide at sunset to rest and sleep. That is also the occasion for Dante to dream.

Finally, the most amazing kinematics is that concerning the journey through Paradise, in which Dante makes the ascent from the first heaven to the vision of God in just twenty-four hours: we are in the presence of a violation of every possible physical law, regardless of the estimate of the extension of Paradise, whether limited to the Solar System or as wide as the boundaries of the known Universe. A space–time journey, therefore, unquestionably unique and improbable.

Marco Polo's journey towards the heart of Asia also begins in Jerusalem, where the Polos, who had set off from Venice, stopped to get oil for the lamps of the Holy Sepulchre to take as a tribute to Kublai Khan. We are in April 1272, and more than three years will pass before they reach the Catai, that is, northern China, where the great emperor of the Mongols resides: a very long period, spent crossing endless deserts, scaling very high mountains, riding over vast and desolate prairies, travelling far and wide along the ancient roads of the caravans, for a total estimated distance of about 15,000 kms, also taking into account, in this calculation, detours or erroneous itineraries. Roads covered on foot, on horseback, on the back of mules and camels.

We know that, when he arrived in China, Marco Polo lived mainly in Cambaluc, today's Beijing, the city that the Turks called Khanbaliq and the Chinese called Yen-ching, after which he lived for a long time in Hangzhou, a fairy-tale city on the water, full of bridges and canals just like his beloved Venice. He earned the trust of the emperor of the Mongols and traveled as an ambassador throughout all the lands of his kingdom, visiting the steppes of Mongolia, the forests of southern China, the thousands of cities of that millenary civilization, covering tens of thousands of kilometers each year. Finally, after seventeen years spent in the magical Orient, he would return to Venice, facing the perilous waters of the Asian oceans, visiting Malaysia, Indonesia, India, Java and the Arabian Peninsula, crossing the Arabian deserts and the waters of the Mediterranean, in a journey almost 20,000 kms long. He would see the dome of San Marco again in November of the Anno Domini 1295, twenty-four years after his departure.

If we would turn to Dante for an account of the wondrous things seen in his dreamlike journey through unknown and surprising realms, to enjoy the exotic sights of distant and magical places, we will also have to appeal to Marco and read his *Milione*. In the following, we will ideally synchronize the temporal duration of the two journeys, obviously a difficult, but not impossible, task, especially if you work with imagination. What follows are, therefore, the stages of our personal journey in the footsteps of the two great travellers.

The Shadow of an Ancient Master

The *Commedia* was born under the sign of Brunetto Latini, a notary by profession and family tradition, as well as the most representative intellectual of Florentine Guelphism, presented in the round by Dante in Canto XV of the *Inferno*. Although Dante has no qualms about exposing the man's weaknesses, placing him in the circle of the sodomites, he nevertheless addresses Brunetto with unusual affection: he calls him "voi" and honours him with the title of "ser," speaking to him with sweet words, full of gratitude and attention.

An evident reference to Brunetto is hidden, in fact, in the incipit of the *Commedia*, long before Canto XV, and the few references we will make here will help us to understand the great influence exerted upon Dante by this Florentine notary.

First of all, it is enough to read the first pages of the *Tesoretto*, Brunetto's famous and delightful work, which is a real encyclopaedic compendium of the time, written in a picturesque language that drew half from Italian and half from French. Brunetto was so proud of his text that he did not hesitate to recommend it to Dante, even in the flames of Hell: "My Treasury – may that commend itself. In that, I still live on. I ask no more." Well, in the *Tesoretto*, Brunetto recounts how, on his way back from the diplomatic mission conducted on behalf of the Commune of Florence to Alfonso X of Castile, he was informed by a student "who came from Bologna and without telling lies was very wise and brave" of the defeat of the Florentine Guelphs at Montaperti on September 4, 1260, and of their expulsion from Florence; because of the great sorrow he felt, Brunetto lost "the great path and went to cross a monstrous forest."

Dante's allusion to the exile of Brunetto, who remained in France for six years, allows us to glimpse, behind the Poet's better known existential bewilderment, his confrontation with the heavy reality of internal political conflicts that inflicted both his and Brunetto's exiles upon them. The *Commedia* was initially conceived, therefore, as the story of a journey among the dead to try to save the living, in particular, the Florentines, who were undergoing a dramatic internal crisis, accompanied by endless social tensions and all of the harmful effects of an economic development that erased the ancient municipal values. But, later on, this poem conceived in favour of Florence paradoxically turns into the most bitterly and violently anti-Florentine book ever written.

As we well know, the journey begins in a "dark forest," crosses a "livid swamp" and continues in the gloom of a world without sky or light: placing a clear echo of the *Tesoretto* by the most illustrious Guelph intellectual in Florence at the beginning of his *Comedy* allows Dante to declare himself the man's political and intellectual heir. And when, in the circle of the sodomites, he finds Brunetto, who calls him "my dearest son," Dante himself recalls his "dear, good father," declares himself his pupil and places himself as the true interpreter of the Guelph communal tradition. Brunetto issues him with a "license" as a citizen of integrity, faithful to the principles of the Florentine community, condemned to exile for "having acted well", one who will have to resist the dangerous demands of the white and black Guelphs who " will seek to set their teeth" for him, because he does not belong to any faction and is disbanded like him.

> **Round Earth or Flat Earth?**
>
> The above question, which still continues to fascinate some credulous contemporaries, was, in the past, the subject of a heated debate among logic, theology and the acute astronomical observations of travelers. And this was the only way it could be, because, in ancient times, it was certainly not possible to refer to images of the Earth taken from artificial satellites or even from the Moon to show the most stubborn Flat Earth fellow the beautiful sphericity of our planet. So, let's take a look at the arguments put forward in the past in favor of the Earth's roundness and the perplexities raised by those who remained skeptical. Because of the intriguing philosophical and geometrical nature of the question, the first to pose the question of the Earth's sphericity were, not surprisingly. the Greeks, in particular, Aristotle, who, in his book *De coelo*, argues in favor of the round shape, noting that, to travelers moving southward, the southern constellations appear to rise higher than the horizon, which could not happen if the Earth were flat. He also astutely notes that the edge of the shadow cast by the Earth on the Moon during the partial phase of a lunar eclipse is always circular, regardless of how high the Moon is on the horizon: only a sphere always casts a circular shadow in all directions, while a flat disk instead casts an elliptical shadow in most directions. Here, according to Aristotle, is the proof that the Earth is spherical and, by similarity, the Moon too.
>
> It was then, in the third century B.C., that Eratosthenes became the first to ingeniously estimate the Earth's radius with the famous parallax effect linked to the wells of Alexandria and Syene, arriving at a value incredibly close to today's estimates made by quite different means. In spite of the stringent Aristotelian logic, there were still many skeptics, who persisted for the most disparate and bizarre reasons. For example, at the dawn of Christian theology, there was much debate about the existence of inhabitants at the antipodes of the then-known world, regions considered distant, separated by the equatorial ocean, and impassable. If the Earth is round, the early theologians reasoned, how can this fact be reconciled with the Christian vision of humanity descending from a single couple and saved by one Christ?
>
> It is curious to note that, for a true traveler, such as Marco Polo, the question has little or no relevance, while for Dante Alighieri, a faithful interpreter of scholastic philosophy, the Earth is strictly spherical, as are the spherical skies that surround it. Of course, there is the embarrassing puzzle of having Lucifer, placed at the center of the Earth, also being the center of the Empireo, but Dante will solve this problem in a surprising way...

But there is much more than this political legacy between the two. The Universe described by Dante is, in fact, the one present in Latini's *Tesoretto*: in one of the chapters, Brunetto describes how the Earth is round and how it would be possible to prove it in a simple way. According to the Florentine notary, in fact, it would be enough for a horseman to always head west to see him then pop up in the east! The geometry of the Earth is therefore not flat, but is, on the contrary, one of spheres, circles, in other words, the geometry of perfection.

Behind the Scenes

Understanding the genesis of a book as peculiar as the *Commedia* is a decidedly interesting exercise: perhaps the first Cantos of the *Inferno*, up to the city of Dite, were written when Dante was still living in Florence, before March 10, 1302, when the chief magistrate of Florence condemned him, in absentia, to be burned at the stake and his house destroyed, while he was in Rome on a diplomatic mission to the hated Pope Boniface VIII. These first Canti present formal, structural and content-specific characteristics that distinguish them from the subsequent ones, especially starting from Canto X (with Farinata degli Uberti), which represents a considerable leap in quality.

When Dante decided to continue his design of a divine poem, he staged a "Florentine" Inferno: until the Canto of Farinata, Dante met, among his contemporaries, only Florentine damned, starting with Ciacco, the first to speak of Florence as a "divided city," an evanescent figure, especially from the political point of view, who allowed Dante to take sides against the moral degradation of the aristocracy and the magnates.

During the years of repentance, of detachment from his Ghibelline friends and the attempt to return to Florence thanks to Corso Donati and the Black Guelphs between 1306 and 1310, and perhaps during his stay in Lunigiana in the second half of 1306, Dante recovered the materials left in Florence and resumed writing the *Inferno*, which he concluded at the end of 1308 or early 1309, during the last period of his stay in Lucca, and published perhaps in the second half of 1314.

He wants to show his loyalty as a Florentine Guelph and dispel, now that he is in exile, his reputation as a traitor. In the *Inferno*, there is a strong link between the *Commedia* and Dante's aspiration to return to his homeland. The poem does indeed speak of the destiny of humankind from an eschatological perspective, but it also closely examines events linked to Florence and current events, with characters who have recently died or are still alive. An instant book in which the oscillations that run through the life of the exile are evident. A book written with posterity in mind, but addressed to a proximate audience, essentially Florentine, with allusions to recent events. And Dante, who certainly could not foresee that he would be committed to it for the rest of his life, aimed it at a restricted and interested public, with direct political messages and hot air.

During the three years of his wandering, through Lunigiana, Casentino and Lucca, Dante tries to get in tune with the Black Guelphs and Corso's faction. This is also why in none of the Cantos preceding Canto X is Farinata degli Uberti mentioned concerning Dante's expulsion and exile. In the *Inferno*, Dante never talks about the causes and responsibilities of his condemnation, because he is trying to get closer to his erstwhile political opponent, Corso Donati. On the other hand, Dante is an exiled character, with severe difficulties, trying to return to his homeland without any knowledge of why he was exiled from Florence. Throughout the *Commedia*, Dante is silent about the decisive years of his life, those from the priory to the banishment. And how could he have spoken of them? He should have denounced the responsibilities of Corso and his side, whose support he now sought. How could an exiled White Guelph in search of amnesty bring attention to the actions of the Black Guelphs? Silence, therefore, on Corso, who is mentioned in the *Purgatorio* by his brother Forese as "who's most to blame" for Dante's exile, but only after he is already dead and after Dante's hopes of returning home with the Black Guelphs have completely vanished. The journey of Dante, and of all humanity, towards salvation in God thus begins in the fire of the Florentine civil war and through an exile whose reasons are to be sought only in the folds of a fierce political and personal vendetta.

In the Heart of Tartary

Medieval Asia presents itself to our eyes as a treasure chest, and so must have appeared to Marco Polo as well. In writing *Il Milione*, the great Venetian traveller was keen to show us not only all that he had seen with his own eyes, but also that great web of secular encounters, clashes, battles and conquests of great commanders that then gave life to an interrupted historical flood, which, starting from the early thirteenth century, would shape the following two centuries. With great historical wisdom and the gift of storytelling that distinguishes him, Marco Polo will introduce us to the extraordinary history of the Mongol people, presenting us, in the pages of *Il Milione*, with a crowd of characters, places, shadows, echoes and flashes of that ancient civilization. So, let's follow him in his tale.

While, in Florence, Guelphs and Ghibellines were struggling in a fight to the death, from the distant steppes of Asia came a people of terrifying warriors. In the West, they were called Tartars, as if they came directly from the dark and gloomy reality of the Underworld: the real name of that group of demons was the Tatars, but the association with the infernal places was too strong to resist the temptation to distort the name.

They came from the heart of great Asia, from the deserts lashed by icy wind and the sand that submerged everything, in other words, from a merciless land: in winter, the temperature often dropped to minus 40 degrees centigrade, while in summer, together with the suffocating heat, there was no lack of thunderstorms of unprecedented violence, accompanied by lightning, hail and lashing rain. All around, there was the immensity of nowhere, the infinite expanse of the plains and, in the distance, the perennially snow-capped peaks of the mountains. This was the realm of gray wolves and leopards, of wild horses and prairie dogs, of hawks and mouflons; on the snows of Tibet roamed heavy yaks; in the Gobi Desert, at sunset, the Bactrian camels stretched their shadows over the golden sand dunes.

In the *Secret History of the Mongols*, it was written that those people were descended from "Blue Wolf" and "Candid Fawn," two divine creatures who came from the Great Sea and who had decided to stop in the lands of the wind to graze their herds and give life to that glorious race. Giovanni da Pian del Carpine, the papal envoy who, in 1245, was among the first Europeans to visit those places and people, described their physical features in detail:

> *They are very different from all other people: their faces, between the eyes and cheeks, are much broader than other men's, and the cheeks protrude strongly above the jaws. Their noses are small and flattened, their eyes also small, and their eyelids are raised to the point of touching their eyebrows. They have narrow hips and average height. The beard is scarce but some have an upper lip and cheeks scattered with a few hairs that they often do not use to shave. Their feet are smaller than normal.*

They had been nomads since the dawn of time, divided into clans who were perpetually fighting each other to enlarge their areas of pasture, for the theft of livestock, for offence towards women or for the pure spirit of oppression. William of Rubruck, another Franciscan friar who wandered the

Asian steppes on a diplomatic mission in the 1200s, noted that "they have no permanent residence and never know where they will be the next day." They moved through the infinite spaces of the steppes, carrying their yurts, the typical round tents erected on a light wooden frame and covered with leather carpets sprinkled with grease. They lived mostly on what little those desolate lands offered, the meat of the mutton or the fermented milk of the mares, sometimes raiding the flourishing territory of their Chinese neighbors. In an attempt to limit the damage caused by these marauders, the Chinese military commanders often cleverly exploited the discord that swirled among them, making sure to set them against each other by offering honorary titles to some clan leaders, but not to others.

"The most prodigious adventure the world has ever known," as an orientalist of the calibre of Paul Pelliot defined it, is that which begins with the arrival on the scene of Temujin, the strong, cunning, violent and unscrupulous young Mongolian prince who came into the world clutching a lump of blood in the palm of his hand. He was the son of a respected leader of the Borjigin tribe, the "grey-eyed Mongols": one of his ancestors had defeated the Chinese army in battle, but, over time, his people had become impoverished and were harassed by other tribes. Temujin was power-hungry, killing his half-brother and best friend one after the other, as well as entire rival groups and anyone else who dared oppose him in his rise to power. His dream was to put an end to the tribes' infighting, and to unite all those unruly nomadic families under his command to form a single Mongolian nation, restoring respect for the ancient rulers of the steppe. To this end, he ruthlessly destroyed the populations that were openly against him—the Keraites, the Naimans and the Merkites—becoming, at the beginning of the 1200s, the absolute ruler of all the lands where the Mongols lived. In a great gathering organized in the Year of the Tiger (1206), at the source of the Onon River, amidst a great waving of flags and shouts that rose to the sky, that young leader was proclaimed Genghis Khan, the king of kings, the leader of all Mongols.

"It is incredible the multitude of Tartars who went to join him," writes Marco Polo in *Il Milione*, "and when Cinghiscan saw himself surrounded by so many valiant men, he provided them with bows and other weapons and set out to conquer other countries." In the fifteen years following his appointment, at the head of ferocious hordes on horseback, he put Asia to fire and sword and upset the balance of the world, creating a single, immense Mongol Empire, extending from the Sea of Japan to Poland, from the steppes of Siberia to the rich plains of China.

His military strength lay entirely in the cavalry and in a wise war strategy, as described, once again, by Giovanni da Pian del Carpine:

> *In front of the enemy they oppose an army of prisoners and slaves and send armies of the strongest men to the right and left, so that the opponents cannot see them. In this way they manage to surround the enemies and only then does the real fighting begin. [...] If the adversaries prove to be so valiant that they manage to open up a passage for retreat, the Tatars favour them but, as soon as they begin to flee [...], they chase them and kill more of them on the run than they could in combat. [...] If they can avoid it, the Mongols shun hand-to-hand combat, but wound horses and men from afar with their arrows, and when the latter are decimated they throw themselves into the fray.*

The savage horde would stop at nothing, not even in the face of the worst weather, for the cold of their plains had made them tough and invincible; they were capable of riding at insane speeds; they were valiant, bold, blindly obedient to their commander and, above all, bloodthirsty. For them, the only life worth living was one of perpetual motion under the stars and through the gusts of cold wind; the sedentary life of the people of the fields or cities was the sordid existence of slaves and inferior beings. In fact, when they conquered North China, the commanders discussed at length whether or not to leave alive the people of those areas so miserably attached to the land.

The Noble Castle

One of the most significant passages of the *Comedy* is certainly Canto IV of the *Inferno*, in which Dante, having awakened after passage of the river Acheron, finds himself on the edge of the infernal "valley of the abyss," animated by the din of endless lamentations: this is Limbo, where the souls of the unbaptized are to be found, including those who lived before the coming of Christ. As he continues with Virgil among that multitude of spirits, Dante sees, far away, a light that illuminates both the darkness and a place inhabited by honorable people who have acquired great fame and high merits in earthly life.

These are the most famous poets of antiquity, Homer, Horace, Ovid and Lucan, who pay homage to Virgil and grant Dante (hurray for modesty!) the highest honor of joining them: "I was the sixth among such great minds."

The group of six poets, discussing poetry and other elevated themes, moves towards the light that can be seen in the distance, finally arriving at "a noble fort," surrounded by seven walls (the liberal arts) and defended by a "lovely stream" (eloquence), which the poets cross on foot as if it were made of beaten earth. They then enter a large green meadow, where various chosen spirits are gathered together: they all have stern looks, an authoritative bearing, and speak in low and gentle voices, surrounded by an aura of austere serenity. In the darkness of Hell, then, here is the luminous oasis of an aristocracy of knowledge, a place that is a veritable eulogy to intelligence.

Along with the heroes and great figures linked, in large part, to the legend and history of Rome (Electra, Hector and Aeneas, Caesar "with his hawk-like eyes," Lucretia, Camilla, and many others), in that group, Dante sees, with reverence and admiration, the "family of philosophers" so dear to him: all of the great Greek philosophers—Democritus, Socrates, Plato and, above all, Aristotle, "the master of all those who think and know"—along with great scientists such as Euclid and Hippocrates. In the noble castle, Avicenna and Averroes, two Muslims, also have their place! But what do these two singular wise men of the Arab Middle Ages have to do with Dante?

The Muslim philosopher and physician Abū 'Alī Ibn Sīnā, called Avicenna in the West, was born in 980 in the Persian city of Bukhara, in what is now Uzbekistan, the same city where, two centuries later, Marco Polo's father and uncle lived for three years before leaving on their first journey to the court of Kublai Khan. The Latin version of his treatise *The Canon of Medicine*, a systematic synthesis of the medical doctrines of Hippocrates and Galen and the biology of Aristotle, was also widely studied in the West. Avicenna also wrote philosophical encyclopaedias, including on such topics as music theory, mathematics, geometry, astronomy, and the natural sciences. One of these, *The Book of Healing*, was partially translated into Latin in the Middle Ages. Avicenna's doctrine, therefore, had considerable influence on medieval thought, particularly because of the common neo-Platonic inspiration of St. Augustine's philosophy.

It is therefore not surprising that Avicenna's philosophy is well present in the cosmological vision illustrated by the *Comedy*. In fact, according to the Persian philosopher, the forms of all things have their genesis in God, who imprints them on matter, uncreated and eternal, by means of emanations in degrading order: the intelligence of the first celestial sphere, which proceeds from the absolute unity of God, the intelligence of the second sphere, and so on, up to the intelligence of the lunar sphere, from which emanates the agent intellect, which is the true cause of the transformations that occur in the sublunar world.

More complex, but also more subterranean and uncertain, is Dante's link with the other great Arabic-speaking philosopher, the Spaniard Ibn Rushd, Latinized as Averroes, also a philosopher, jurist, doctor and astronomer. Believed by some to be a martyr for free thought, siding against Islamic fundamentalism—as told, for example, in the film *The Destiny* by Youssef Chahine (1997)—in reality, Averroes was one of the greatest interpreters of Aristotle's writings, on which he wrote his famous *Commentaries*. In fact, it is thanks to Averroes that the Aristotelian tradition, which had been almost completely forgotten, was recovered and profoundly re-evaluated in Western culture.

In his most significant philosophical work, with the recursively oxymoronic title *The Incoherence of the Incoherence of Philosophers*, he raised his voice against other Arab thinkers who were highly critical of Aristotle because they claimed he was in contradiction with Islam. In contrast, Averroes claimed that truth could be reached both through religion and speculative philosophy, a thesis close to the one adopted centuries later by Galileo Galilei to save himself from the stake of the Inquisition. The modernity of Averroes is found, with all its stainless vitality, in a series of theses that spread in the West in both Hebrew and Latin versions, and were soon judged heretical: the independence of the truths of reason, proper to philosophy, from the truths of faith; the eternity of matter and the world; the denial of the immortality of the individual soul.

Within Christianity, Averroism was seriously opposed as a heretical doctrine. The subject of discord was Averroes' afore-mentioned denial of the immortality of the individual soul and his theorization of a "double truth": the philosophical truth, known to a select few, and the truth of faith, made up of myths and religious practices that were good for the people. The greatest follower of Averroes among Christians was Sigieri of Brabant, a professor at the Faculty of Arts in Paris from 1266 to 1276, who argued that Aristotelian doctrines did not coincide with the teaching of the Church. For this reason, Sigieri was accused of Averroism by the greatest theologians of Dante's time, such as St. Albert the Great, St. Bonaventure of Bagnoregio and St. Thomas Aquinas: suspected of heresy, he was assassinated in 1282 or 1283 under unclear circumstances in Orvieto, where he had gone to clear himself of the accusations before Pope Martin IV, according to what Dante tells us in his poem "Il Fiore" (XCII, 9–11).

In light of all this, Dante's choice in *Paradise* to have Sigieri praised by his adversary Saint Thomas is quite singular:

This is the everlasting light of Siger,
whose lectures, given in Straw Alleyway,
argued for truths that won him envious hate.

Paradiso, X

As we have already mentioned, the "Straw Alleyway" is rue de Fouarre, still located near the Sorbonne in Paris, the place where Sigieri, who, on a philosophical and theological level, is the most enigmatic figure in the *Comedy*, taught. Sigieri argues that freedom is to be sought in human reason, in the intellect that has the power to evaluate any external cause. Sigieri further asserts that both human action and physical action are governed by a single natural law. Unlike the Augustinian school, which attributes human progress to the intervention of God, Sigieri places the intellect at the center, and therefore considers human progress to be intrinsically linked to education and the deepening of thought.

In this sense, one might find a consonance of Sigieri with the fundamental rhythms of the *Commedia*, which frames the motion of Dante's will within the broader motion of the entire physical world. In Dante, we see the triumph of faith in the absolute conciliation between science and philosophy on the one hand, and Christian faith on the other. Sigieri is placed in Paradise as the symbol of a philosophy that is completely autonomous from faith, yet convergent with it. Dante wants to separate the two spheres, that of philosophy, relative to earthly beatitude, and that of revelation, relative to heavenly [?] beatitude, and thus make them independent.

It is widely believed that Dante became acquainted with the work of Averroes at a young age, above all thanks to his friendship with the poet and philosopher Guido Cavalcanti, considered by many scholars to be an Averroist, an adherent of the secret sect of the Fedeli d'Amore. Other interpreters deny Dante's tacit Averroism, which, in their opinion, would be contradicted by the entire structure of the *Convivio* and the *Commedia*. And they maintain that Sigieri is placed in Paradise to denounce his unjust killing when the philosopher of Brabant went to pay reverent homage to the pope to defend and clarify his position. However, this interpretation would imply a certain naiveté and incompetence on Dante's part in understanding the heretical consequences of Sigieri's philosophy.

Perhaps those who maintain that Dante was neither an Averroist, nor a Thomist, nor exclusively Aristotelian, nor merely Neoplatonic, nor purely Augustinian are right. What is certain, however, is that he loved philosophy as a faithful Christian who had studied at the Faculty of Arts in Bologna with Michele Scoto, Sigieri and Martino di Dacia, leaders of the Latin Averroists, after having become acquainted with the doctrine of Averroes from his dialogue with his friend Giudo Cavalcanti.

That the Poet was not strictly an Averroist is shown by his philosophical passion for Albertus Magnus. The theologian of Cologne, known as Doctor Universalis, was a teacher of Thomas Aquinas and a great promoter of the spread of Aristotelian philosophy in the West. But he was also a great scholar of the natural world, very attentive to the observation of natural phenomena. His program was to "render intelligible to the Latins" the Aristotelian philosophy, through, among other things, free paraphrases of the works of Aristotle and his main commentators, including Avicenna. His exposition of Aristotelian philosophy, with a strong accentuation of Platonic themes, exerted a great influence on Dante, as did his clear separation between peripatetic philosophy and theology, in the awareness that "physical principles do not agree with theological principles." Alberto wrote numerous scientific treatises (*De animalibus*, *De vegetalibus*, De mineralibus, etc.), in which prevails the taste for the direct observation of nature so present in Dante, who quotes him three times in the *Convivio*. But the presence of Albert is constant throughout Dante's meditation, which proceeds from the wisdom of the ancients to find, with Aristotle and beyond Aristotle, the reasons for the laws of the world and to reach the free "happiness of the intellect" together with the Christian beatitudo.

Albert's place can therefore only be in Paradise, next to his disciple Thomas:

This one, who here is nearest on my right,
was master to me, and a brother, too –
Albert of Köln. I'm Thomas Aquinas.

Paradiso, X

Dante's adherence to the "philosophical family" was a sign of his participation in the great Scholasticism of the thirteenth century: he mastered both the corpus of Aristotle, now fully known in its Latin version, and the other philosophical and scientific writings connected with Aristotle's fortunes. The pages of Thomas Aquinas—the Dominican friar who, despite his humility, was sought after by all of the most important centers of study of the time, including the Sorbonne in Paris—were constantly rattling around in his head, with his reckless way of reconciling faith and reason, of always moving along the edge of a precipice, like a tightrope walker. The questions that had obsessed Thomas were also his, because, reasoning according to logic and honesty, he always arrived at disruptive conclusions, at the limits of heresy: for example, was the world eternal? Had it always existed? If so, it was obvious that there was no longer a need for a creator God, but that would put a damper on what was written in the Bible. But, if God had created the world from nothing, the problem changed shape, because it would then be necessary to decide whether this "nothing" came before God… And then, why would God decide to create the world? Questions upon questions, and that devil of a Dominican had always found a way to get away with it: he claimed that faith and reason were compatible, but, cleverly, he had established that there was a useful hierarchy between the two. And, in case of doubt, it was always faith that prevailed. A lesson to treasure.

Pax Mongolica

In the first decades of the 1200 s, for many European peoples, terror came from the Far East, with increasing news, although fragmentary and imprecise, of a people of ferocious and invincible warriors who destroyed everything in their path. It was also rumored that they were the descendants of the biblical monsters Gog and Magog, who were directly stirred up by Satan to kill all Christians. The chronicles written by Chinese officials, Arab merchants, Persian dignitaries and Christian missionaries give us a staggering number of dead—several million men, women and children, without distinction as to sex or age—who had the misfortune of falling into the path of Genghis Khan.

The ritual was always the same: after defeating their enemies, these ferocious warriors dedicated themselves to the sacking of cities and the systematic killing of civilians. Old people, men and women were made to sit on the ground and undressed so that their clothes, considered war booty, would not be stained with blood; then, the executioners, armed with double-edged axes, would kill all of the prisoners without pity. Those who remained alive regretted not having died under the blows of the axes, especially those young women or men who were considered fit for the role of slaves, who suffered the mutilation of their noses and disfigurement of their faces. Enemy princes were skinned, quartered, burned with red-hot brands or molten iron poured into their ears, stripped naked and turned into hunting prey to excite the troops' sense of ferocity even more.

In spite of all these horrors, it is nevertheless impossible to ignore the paradox that accompanies the figure of Genghis Khan: his ferocity was, in fact, followed by a policy of extraordinary tolerance and sincere curiosity about the practices and customs of the peoples he conquered. When he arrived in Karakorum, the Mongolian capital, the Franciscan friar Giovanni da Pian del Carpine was amazed to find Nestorian Christians, Muslims, Buddhists and shamans, all serenely at peace among themselves. All faiths had the same dignity—Genghis Khan had decreed—as long as their ministers accepted the authority of the emperor, and any violation was punished with death, because it was an offense against the laws of heaven.

Genghis Khan's sweeping ride had led to the construction of the greatest empire of all time, eclipsing the conquests of Alexander the Great and those of ancient Rome. Different peoples found themselves under a single body of laws, called yasaq, very rudimentary but still a system of laws: among its articles, there was room for prohibitions dictated by common sense ("It is forbidden to bathe or wash clothes in running water when it thunders"), as well as capital punishments for various reprobates ("Spies, false witnesses, men devoted to infamous vices and sorcerers are condemned to capital punishment").

When he died on August 25, 1227, Genghis Khan's body was transported on a cart to the slopes of Burkhan Khaldun, in the heart of Mongolia, and buried there in a secret location. In order not to deny his sinister reputation even in death, the guards escorting the coffin killed everyone they met along the way, men and animals, so that they could join the army of those who were going to serve the great Khan in the afterlife. The entire area around the burial was restricted for hundreds of square kilometers and trampled by hundreds of horses to erase any trace that might be useful for its location.

The immense empire he created was divided among his four sons: Giutsci, Giagatai, Ogodei and Tului. Giutsci got the western lands, together with Siberia and Turkestan; Giagatai got the ancient territory of Qara Khitai; Ogodei got the territories between Balkash and Altai; Tului got the Mongolian highlands. Strong frictions immediately arose among the four brothers, and the one who prevailed in the end was Ogodei, the closest in temperament to his deceased father. He became Great Khan in 1229, and, having come to power, reorganised the army and sought to consolidate the previous conquests of central and northern China and Persia, crushing all outbreaks of resistance. In 1235, he entrusted his nephew Batu with command of the army and ordered him to conquer the grazing lands west of the Volga, which, at the time, were occupied by the Turkic Cumans. After they were done with the Cumans, the Golden Horde of the new Mongols, under Batu's command, overran the Bulgarians, Hungarians, and Poles, whereupon they returned to the endless meadows of the Mongol steppes.

Ogodei died in 1241. After a few years of regency by his widow, Torogene, Ogodei's son Guyuk succeeded his father, but he died prematurely, due, in part, to his heavy drinking. A few years later, the Great Khan Mongka ascended to the throne. His eight-year reign was a period of relative tranquility, and, upon his death in 1259, he was succeeded by his brother Kublai, the Great Khan at whose court the Polos lived for seventeen years.

In spite of all the genocides, mass deportations and destruction of cities razed to the ground and rebuilt from scratch, the Mongol Empire, created by Genghis Khan and consolidated by his successors, appeared solid and peaceful, with people of different lineages, languages and religions living harmoniously together under the fair, but inflexible, pax mongolica. "A virgin with a plate of gold could wander undisturbed from one corner of the Empire to another," it was said. Moreover, the conquest of Persia and the Mongolian policy that did not impose duties on goods, allowing them to travel freely along the Silk Road, established, for the first time, a concrete contact between East and West, a connection that opened the doors to the journey of a thousand merchants and the passage to the East of Marco Polo.

Francesca and the Others, or Love and Sin

In the *Comedy*, the first woman to be named is Beatrice, with verses that reveal how, for Dante, the real spring that moves the world is love:

Her eyes were shining brighter than the stars
Inferno, II.

According to the courtly literature of the "dolce stil novo" in which he was trained, for Dante, love involves us and moves us, but, above all, improves us, because it elevates us to heaven. Beatrice is therefore the one who announces salvation. She is:

the sun who first in love shone warm
into my heart

Paradiso, III.

She is the woman who has come from heaven to announce a prodigy and who adds her mystical seduction to the divine transcendence: overcome by emotion, Dante whispers that she is "the one who brings my mind to paradise." In the blaze of light and ardour of Paradise, the more Dante ascends towards the higher heavens, the more luminous and ethereal Beatrice becomes, until she vanishes into the candid rose of the blessed:

And she, as far away
as she might seem, smiled and looked down at me

Paradiso, XXXI.

Beatrice, however, is not the only female presence: love is everywhere in the *Comedy* and has a thousand facets, passionate and gentle, desperate and lustful, insidious and miraculous. If Dido, Cleopatra and Helen, together with the unfortunate Francesca, are condemned, by the merciless law of opposites, to be dragged restlessly like twigs by an unstoppable, whirling wind, in Purgatory, we encounter the sweetness of Pia de' Tolomei, murdered by her husband, who caused her to fall from a balcony of their castle in Maremma so that he could remarry:

please, do remember me. I am La Pia.
Siena made me, unmade by Maremma.
And he knows this who, once I wore his ring,
took me in marriage with his own bright gem.'

Purgatorio, V

In Paradise, on the other hand, in addition to Saint Lucia and the Virgin Mary, we find the divine love of Piccarda Donati. But it is obviously the love for Beatrice that reveals itself as the illustrious poetry of intelligence.

However, this is certainly not Dante's only declination of femininity: in front of the "carnal sin made reason bow to their instinctual bent," Dante confesses that " grief and weeping pierce me at the heart." Is it not well known that Dante was a lustful man? Guido Novello da Polenta, his protector and lord of Ravenna, recounts that the Poet spent a night with a "market woman," and that the woman told Guido "she was of little worth, because having had very good beasts under her, she had not ridden more than a mile"; but he adds Dante's prompt reply: "I would also have dropped the ace, but I did not like the dealer."

And the Comedy is also well supplied with courtesans, together with the furies, the sorceresses, the soothsayers of antiquity, and all of those scurrilously depraved women, such as the harlot Taide of Malebolge:

Thais! She's there, the whore, the one who cooed
to her hot panting swain ("Yeees! Good for you?"),
"Angel, a miracle! My thanks indeed!"

Inferno, XVIII.

The encounter in Canto V with the great figures who embody the diseases of love in universal literature, from Helen to Dido, from Paris to Tristan, strikes Dante more than anything else, and in the face of "these chevaliers of old, and noble ladies, pity oppressed me and I was all but lost." Quite different is Dante's intention in presenting the first sinners who were contemporary with him, two people from Romagna, Francesca da Polenta and Paolo Malatesta, who immediately took on the burden of their sins: "…who tinged the world with blood." They are the protagonists of a story that is very well known in Florence. Paolo had been captain of the Comune between 1282 and 1283, and Francesca's father became potestà between July and November 1290, a few years after the lovers' crime, which can be dated to 1285. The crimes committed within the amorous union of Paolo and Francesca, by "the ill perverting us," are well identifiable in the jurisdiction of the time: adultery, because Francesca betrays her husband Giancotto; incest, because the lovers are in-laws; and murder, because the two are killed by the betrayed husband, who will also be damned forever ("Cain's ice awaits the one who quenched our lives").

The two had sinned in lust, but, by submitting reason to "talent," they had also provoked serious social effects: two exponents of the most prestigious feudal nobility had abandoned themselves to illicit behaviour, a sign, for Dante, of the degradation of his times and the disappearance of those traditional values proper to the aristocratic class that should have acted as an example for the whole of society.

And yet, Dante does not remain insensitive to their love affair: he cannot, both because of his own personal inclination and because of his great familiarity with love poetry. The verses "Amor, ch'al cor gentil ratto s'apprende"—which echoes the line from the Vita Nuova "Amore e 'l cor gentil sono una cosa"—and "Amor, ch'a nullo amato amar perdona" have nothing conventional about them. And it is no accident that Dante feels himself engaged in a question that is both poetic and philosophical:

the how of it – and why –
that Love, in sweetness of such sighing hours,
permitted you to know these doubtful pangs.

Inferno, V

Francesca's answer lies in her reading of the French novel *Lancelot of the Lake* from the Breton cycle of the Round Table and in Galehault (Galahad), Lancelot's friend and companion—"This book was Galehault—panderpenned, the pimp!"—who induced Guinevere to kiss Lancelot in that novel. So it happens, inversely, with Paolo's kiss with Francesca.

In Dante's philosophy of love and that of Stilnovists like Guido Cavalcanti, men recognize themselves through the power of thought and love is a passion that is also present in the sphere of the intellect, not only in the sensitive part of the soul. Guido, as a good Averroist, meant, for example, in *Donna me prega*, that the conjunction between love and intellect lives in both witnessed and imagined forms, in this case, according to Dante, in the forms found in the chivalric romance read by Paolo and Francesca. The experience of love is the immoderate cogitatio of an interior phantasm, Andrea Cappellano wrote in his *De amore*. And to quote a French psychoanalyst and philosopher of our time, Jacques Lacan, "le fantasme fait le plaisir propre au désir." Faced with these phantoms, it is difficult to resist, and Dante, after hearing Francesca, cannot avoid fainting from emotion:

fainted away as though I were to die.
And now I fell as bodies fall, for dead.

Inferno, V

Dante, however, will be saved from lust, because his "amorous thinking" will be resolved in the desired image of Beatrice, which, beyond individual experience, will be translated on the level of eternity and salvation.

Women of the Orient

In contrast to Dante's colorful world of women, so dramatically poetic and often idealized, the universe of women recounted by Marco Polo is as pleasantly human and carnal as can be imagined. In *Il Milione*, there are several chapters featuring amusing accounts of an upside-down family, at least by Western standards and the medieval customs of the time.

The first story, with a distinctly Boccaccio-like flavour, concerns the province of Kamul, situated between two great deserts, the Gobi and another, less extensive one, three days' walk from the first. Marco wants to inform the reader that the inhabitants of the place are all idolaters who live off the produce of the land, but who love to drink and offer drinks to passing travellers: "They are playful and pleasure-loving men who only care about playing, singing, dancing and taking carnal pleasures."

The surprise comes, however, when the subject of the local women comes up! They would give themselves with great joy to the stranger passing through, or rather, when a traveller knocked on the door of a house to ask for accommodation, it was the husband himself who would put all of the women of the house, daughters, sisters and even his wife, at the guest's disposal, provided, of course, that he paid. Indeed, to make his stay even more pleasant, the master of the house usually left the home for two or three days, so that the stranger could share the bed with the woman in great pleasure. This was a genuine custom of the place, and the inhabitants were not at all ashamed of it. The women were "beautiful, cheerful, and sensual."

However, Möngke Khan, leader of the Tartars, nephew of the great Genghis Khan, messed things up. After conquering those lands, he decided to put an end to all of that upside down behavior in the bedrooms, under penalty of severe punishment. To solve the problem of accommodation for foreign travellers, he simply decreed that hotels should be built. For three years, the people of Kamul tried to obey the Khan, but seeing that the land was no longer as fertile as it once was and that the families were afflicted by great misfortunes, they gathered in council and decided to send a large gift to the emperor, begging him to leave their women free to behave according to the customs of their ancestors. It was the ancestors themselves who had inculcated in them the belief that the idols they worshipped would be pleased to bestow happiness and prosperity if the women bestowed pleasure on the wayfarers. "If you are happy with your dishonour, keep it," was the lapidary reply of the Khan, whereupon the old customs, witnessed by Marco Polo himself, came back into vogue in the province of Kamul.

This was certainly not the only curious experience that he had. For example, in the province of Yutian, in north-eastern China, Marco Polo was amused to observe the rather flexible marriage customs: if a husband was away from home for more than twenty days, his wife could proceed quickly to find another husband. There was, however, an interesting and fair principle of reciprocity: the travelling man could choose a bride in every place he visited, in every country where he owned, for example, a shop in the local bazaar. Needless to say, all of this gave rise to a crazy whirlwind of children, cousins, half-brothers and half-sisters, which greatly enriched the genetic heritage of those peoples.

Another singular sexual custom noted, with the usual disenchantment, by Marco Polo concerned the people who lived in the province of ancient Tibet, surrounded by the majestic peaks of the Himalayas. In those lands, there was a strange custom for getting married: no man was willing to take a virgin girl as his wife. A woman who had not had several men was certainly not loved by the gods, and therefore had no value. Given these premises, how can we be surprised that every time a new traveller passed through, the old women of the village would gather together ten, fifteen, twenty girls and offer them to the stranger so that they could lie down with him? Those who were successful remained with the traveller, whose only obligation before leaving was to provide them with a tangible proof of his affection. This usually consisted of a piece of jewellery or a necklace to wear around the neck. In this way, those who had twenty necklaces could prove that they had had at least twenty lovers and, the more jewels there were, ça va sans dire, the more they were put forward as suitable to be taken as wives.

"When they have married them," wrote Marco Polo, "their husbands hold them very dear, because they consider it a great sin to touch another man's wife, and they are very wary of this sin." And, with a touch of irony, he concluded: "In this district young people between sixteen and twenty-four years of age would gladly go"… without specifying which districts!

Monsters, Animals and Giants

In his visionary and fantastic journey, Dante enriched the nefariousness of the *Inferno* with the description of many terrible demons. One of the most dreadful is Geryon, "The beast who soars with needle tail," equipped at its end with a.

> *venomous fork.*
> *The tip was armed like any scorpion's.*
>
> *Inferno*, XVII.

Geryon emerges from the chasm of Malebolge, and is the very symbol of fraud, a sin punished in the last two circles of Hell. Along with this sin, in the infernal chasm, we find a whole other zoology of sins: avarice as personified by Pluto, the Greek god of riches and the underworld, who, in the fourth circle of Hell, cries out in a hen's voice, spouting rambling phrases such as "Popoi Satan, popoi Satan! Alezorul!"; violence in the form of the Minotaur, the monster with a human body and a bull's head ("It cannot walk but skips and hops about"), who, in the seventh circle of the violent ones, as soon as he sees Dante and Virgil, begins biting itself out of rage; and Minos, who knows about sin and, for this reason, has become the demonic judge of souls in Hades, as already detailed in both Homer and Virgil.

Geryon is described as having the face "of an honest man." He is" like a serpent all down his trunk", with "two paws, both hairy to the armpits": a monstrous representation typical of the sculptures and gargoyles of churches in the Middle Ages, a figure that has a human face, the tail of a snake and the legs of a ferocious beast, all symbolic expressions of fraud. But Dante adds other very effective iconic elements:

> *two clawing grabs, and hairy to the armpits,*
> *its back and breast and ribcage all tattooed*
>
> *Inferno*, XVII.

That is, his skin was tattooed with weaves and small round shields that symbolize the fabric of deceit that the fraudulent uses for his defence. But they are tattoos that recall oriental fabrics:

> *with knot designs and spinning little whorls.*
> *No Turk or Tartar wove a finer drape,*
> *more many-coloured in its pile or tuft.*
>
> *Inferno*, XVII.

Dante had had the opportunity to notice the oriental style at the court of Cangrande della Scala, who was, for his part, so fascinated by the fabrics and artefacts brought by the Polo family and described in *Il Milione* that he had himself buried with precious Chinese canvases.

The flat parts and those in relief described by Dante, some subdued, others superimposed, are drawn and coloured in the style of embroidered fabrics from the East or like the canvases woven by Arachne (or Aragne, as Dante calls her), the daughter of the dyer Idmone described by Ovid in the *Metamorphoses*, very skilled at weaving and so daring as to challenge the goddess Athena, a master in the art. For the challenge, Arachne chose the theme of the loves of the Gods: her work, so perfect in describing the amorous wiles of the Gods, angered Athena, who destroyed the cloth, hit Arachne with her bobbin and turned her into a spider, forcing her to spin for the rest of her life.

Dante will remember Arachne again in the second circle of Purgatory, where the proud are punished by being shown thirteen ancient stories of pride carved into a large frame. The "Mad Arachne" is portrayed.

> *half-turned to spider and the work in shreds*
> *which, once attempted, brought you so much harm.*
>
> *Purgatorio*, XII.

In other words, she is fixed in her monstrous transformation into a spider among the torn shreds of the web destroyed by the goddess.

Alongside this exotic image of the "spider-woman," Dante dusts off the memory of the fabrics seen at Cangrande's court, those same fabrics that Marco Polo had helped to introduce to the lord of Verona and throughout Europe: to clothe Geryon with tattoos similar to Chinese garments does not exactly express an appreciation for the exotic customs of the East, but it is a sign that Dante's *curiositas* extends to all aspects of the world, not the least of which is nature.

In the *Comedy*, there are, in fact, gestures towards and glances of profound biological insight regarding the habits of various animals. At least sixty animal species are mentioned. In Canto XXXI, for example, Dante enters the ninth circle of Hell and speaks of the "dreadful giants," "towering here to half their body height," they who "stand within the well around its rim, navel height downwards, all the lot of them." The giants correspond, among humans, to the elephants and whales among animals, and Dante finds a way to reflect on the strange presence in nature of these mammoth animals, comparing it with that of the infernal giants:

If she [Nature], on that account, did not repent
of whales and elephants, to subtle minds
this will seem right, and most intelligent.
For when the powers of working intellect
are wed to strength and absolute illwill,
then humans cannot find a place to hide.

Inferno, XXXI

Verses of surprising ethological finesse, especially since Dante mentions animals outside of the context of metaphor. If what differentiates humans from other animals is "the subject of the mind," the giant is different from a normal human, therefore, only in terms of proportions and perspective.

The giants are traitors, they are losers, but their example retains its value even in the terrible perversion of hell. They remain stuck in the ice, condemned to the immobility of their abnormal dimensions, like "its circling curtain wall, Montereggione [a turreted village near Siena] boasts a crown of towers crowns itself."

And "more than ever death more than ever now in mortal fear" because of the earthquake produced by the movement of the giant Phialtes, Dante sees another giant approaching: Antaeus, son of Neptune and Gaea (the goddess Earth), killed in mythology by Hercules. When he sees him bending over him and Virgil, grabbing them and carrying them in his fist to the icy depths of Hell, he experiences a hallucination:

Just as the Garisenda tower, when viewed
beneath its leaning side, appears to fall
if any floating cloud should pass behind

Inferno, XXXI

It is as if he could see the Bologna tower of the Garisenda looming above him, seemingly leaning in the direction of the approaching clouds.

Dante has previously met other giants. The first was Nembrod, who pronounces strange words—"Raphèl maìamècche zabì almì"—and is apostrophized by Virgil, who invites him to vent his anger by playing the hunting horn on his shoulder. The master explains to Dante that the language of the giant is incomprehensible to all, as he was responsible for the confusion of languages at Babel. But it seems that the words, mispronounced, find their origin in Hebrew, the supposed original language of Nembrod, according to these correspondences: raphaim = gigantes, man = quid est hoc?, amalech = populus labens, zabulon = habitaculum, alma = sancta; thus, they could be translated as: "Giants! What is this? People lapping at the holy habitation." The second giant was the recollected Fialte, who, together with Oto, had tried to climb to heaven by placing the two mountains Ossa and Pelio on top of each other: Fialte is tightly bound by a chain and struggles angrily. Dante expresses the desire to see Briareo, whom the myth described as having a hundred arms, but Virgil declares that his appearance is identical to that of Phialte, and therefore entirely human. Two other giants, Tizio and Tifeo, also present in the infernal well, are mentioned but not seen. The giants appear again, as an example of punished pride, at the exit of the first frame in Canto XII of *Purgatory*, immediately after Lucifer: the sculpture on the floor of the frame depicts Briareus struck down by divine thunderbolts, the other giants exterminated after the battle of Flegra, and Nembrod at the foot of the tower of Babel, lost and confused.

But let's go back to the wealth of exotic animals, somehow intermingled with infernal monstrosities. In Canto XXIV of the *Inferno*, in which the thieves of the seventh bolgia are described, there appears a large crowd of snakes:

of such diverse kinds
the memory drains the very blood from me.
Let Libya boast – for all her sand – no more!

Inferno, XXIV.

On earth, Libya is the privileged place of the most terrible serpents, but in Hell, there are worse. Dante chooses five of the snakes described by Lucan in the *Farsaglia*, that is, in the account he gives of the Romans' march through the Libyan desert: chelidri (amphibious snakes), iaculi (snakes that throw themselves from trees onto men), faree (in Dante, feminine, from Lucan's masculine phareas, snakes that proceed straight ahead, furrowing the ground with their tails), cencri (snakes of various colours) and amphisibene (who have a second head instead of a tail). This is a further expression of that biological attention and that *curiositas conoscitiva* that never abandons Dante.

Hell is populated by many other demonic monsters, such as the aforementioned "Charon the demon, with his hot-coal eyes," the helmsman who, in Canto III, ferries Dante and Virgil across the river Acheronte. Or, in Canto VI, the three-headed dog Cerberus, "weird and monstrously cruel, barks from his triple throats in cur-like yowls," in the circle where the vice of gluttony is punished. The figure of Cerberus, "that reptile," a biblical term that represents the torment that will gnaw at the damned like an eternal worm and that Dante will also attribute to Lucifer, expresses all of the violence of an infernal demon:

His eyes vermilion, beard a greasy black,
his belly broad, his fingers all sharp-nailed,
he mauls and skins, then hacks in four, these souls.
[…].
He stretched his jaws; he showed us all his fangs.
And me? No member in my frame stayed still!

Inferno, VI

A swarm of ten devils also appears, comically, in Canto XXI, which describes, in the fifth bed of the eighth circle of Hell, the punishments of barterers, guilty of having used their public offices to enrich themselves by buying and selling measures, permits, privileges: this is what we would today call the crime of extortion, the same crime of which Dante was accused in the farcical trial that led to his condemnation.

The barterers, totally immersed in boiling pitch, are guarded by the winged and black Malebranche demons, armed with hooked sticks with which they seize and tear to pieces any damned person who tries to emerge from the pitch. Dante introduces them to us with their picturesque and comical names: the leader is Malacoda, who summons to the gathering the decuria that is to escort Virgil and Dante. Thus appear in order Scarmiglione, with his dishevelled hair; Alichino, who is none other than the devil of medieval theater, also known as Hallequin, or, even more familiar to westerners, Harlequin; Calcabrina, also know as the "frost trample" or the "grace stomper"; Cagnazzo, the big dog; Barbariccia, with his disheveled beard; Libicocco, libeccio and scirocco, who together represent the stormy winds; Draghignazzo, the dragon's sneer; Ciriatto, the little "ciro," meaning "swine" in Tuscan; Graffiacane, who scratches like a dog; Farfarello, the imp (in old French farfadet), of whom Leopardi will write in *Dialogo di Malambruno e di Farfarello*; and Rubicante, the rabid (from rabies, "rage"). Their rather comical names together produce a visual and phonic element with an effect of farcical aggression.

Finally, we cannot forget among the infernal monsters, in the deep well of the ninth circle, Lucifer, the prince of devils, the rebel angel, before whom Dante is terrified and petrified ("I neither died nor wholly stayed alive"):

How great a wonder it now seemed to me
to see three faces on a single head!
The forward face was bright vermilion.
The other two attached themselves to that
Along each shoulder on the central point,
and joined together at the crest of hair.

Inferno, XXXIV

The three faces in a single head, joined in a crest similar to that of animals, with three different colours—red, yellowish and black—represent impotence, ignorance and hatred, vices opposed to the three aspects of the divine trinity, power, wisdom and love.

Behind each face there issued two great vanes,
all six proportioned to a fowl like this.
I never saw such size in ocean sails.

Inferno, XXXIV

With his wings, as big as enormous ship sails, Lucifer "fluttered" like a bat, "so that three winds moved from him." The description, which continues for nine tercets, shows Lucifer again "with six eyes weeping, and for three minds dripping weeping and bloody drool," while, with each of his three mouths, he mangled three sinners and, at the same time, scratched with his sharp nails the most serious sinner, in the centre, Judas Iscariot, the traitor of Jesus Christ.

Having arrived at the heart of Hell, twenty-four hours from the beginning of their journey, Dante and Virgil find themselves at the centre of the Earth and, at that point, they reverse their direction. From the time of destruction and sin, which proceeds in the *Inferno* to the point of no return centred on Lucifer, they pass to the eternal time of Purgatory, with a reversal of motion and orientation. The inversion takes on the features of a modern film sequence:

But then, arriving where the thigh bone turns
(the hips extended to their widest there),

my leader, with the utmost stress and strain,
swivelled his head to where his shanks had been
and clutched the pelt like someone on a climb,
so now I thought: 'We're heading back to Hell.'

Inferno, XXXIV

Virgil, along with Dante, clinging to the hairs of Lucifer's pubic bone, turns upside down and comes "out of the hole of a stone" in the opposite hemisphere to that centred on Jerusalem, which instead has the mountain of Purgatory at its centre. This image closes the worst nightmares represented by the infernal monstrosities, worthy of a visionary journey into the darkness of our mind and the obsession of our anxieties.

The Ineffable Priest Gianni

Wherever he went, Marco Polo listened to the local tales and legends and what the merchants, missionaries, soldiers and beggars who passed through saw along their way. The Silk Road was full of rumours from the past, tales of bloody battles or of great kings, stories that changed plots and protagonists every time they passed from mouth to mouth. This is the case of the story of the priest Gianni, who is presented to us by Marco Polo as the ruler of the Tartars until the day they chose Genghis Khan (Cinghiscan) as their sovereign.

When Cinghiscan had gathered enough people to populate the whole earth, he announced that he wanted to conquer most of the world. And he sent some of his messengers to Prester John, the leader of the Tatars at the time, to ask permission to marry one of his daughters. Prester John, upon hearing the ambassadors' message, was deeply offended, and replied: "How is not Cinghiscan ashamed to ask for my daughter in marriage? Does he not remember that he is my subject and my servant? Go back to him and tell him that I would rather see my daughter burned than give her to him in marriage."

War between the two became inevitable. Genghis Khan brought his troops to the great esplanade in front of the enemy camp and, before the battle, summoned Christian and Saracen astrologers into his presence so they could tell him who would emerge victorious. The Saracens could not discern the response of the magical arts; the Christians, on the other hand, contrived to satisfy him. They placed before the Great Khan a piece of green cane, cutting it in two lengthwise, and putting one piece on one side, representing Cinghiscan, and the other piece on the other side, representing Prester John. Then, they said, "Sir, look at these pieces of cane and see which of them bears your name and which the name of your enemy. After we have made our incantations, the piece that dominates the other will indicate the winner." The Christian astrologers then took the Psalter, read certain psalms, and cast their spell. Without anyone touching it, the piece that bore the name of Cinghiscan approached the other and jumped on top of the one representing Prester John. All those present saw the miracle, and Cinghiscan greatly rejoiced.

Two days later, the enemy army was, in fact, routed and the priest Gianni died in the fight. This is what Marco Polo tells us, leaving out, among other things, what happened to Gianni's daughter, the casus belli of the story.

The point is that, in the jumble of legends and beliefs, of names of princes of the Turkish-Mongolian tribes and tales of battles, Marco probably mixed up names and events, making the priest Gianni the direct rival of Genghis Khan, when, in fact, he was and is one of the most elusive and intriguing figures of the medieval world. His is a tasty story, linked to one of the most striking and ingenious historical fakes ever conceived. Let's take a look at the most prominent events.

The first indications of the existence of the priest Gianni occurred in November 1145, when Pope Eugenius III had received the unwelcome news from the Syrian bishop of Antioch, Hugh of Jabala, that the city of Edessa had fallen into Muslim hands. But, to alleviate the distress of that news, the bishop had added that he was aware of a major victory over the Muslims in the land of Persia by a mysterious Christian king: this king was directly descended from one of the three Magi who had paid homage to the infant Jesus. He was said to be a Nestorian ruler, of that sect that had distanced itself from the Church of Rome because of disagreements about the human and divine nature of Christ, but which had spread throughout the vast lands of the East. Otto of Freising, a chronicler of the time, paid particular attention to the Syrian bishop, who spread the legend of a Christian king from the Far East, eager to help Christianity free the Holy Land from the onslaught of Muslim scimitars.

But was there any basis to those reports? Prince Sanjar's Islamized Turks had, in fact, suffered a devastating defeat at Katvan, a town near Samarkand, in 1141, by an army that had come from China under the command of Yelu Dashi: an army composed mostly of Mongols, with some Nestorian soldiers and followers of Confucius. There was therefore no priest Gianni, no man of the Christian faith devoted to war in any quarter of Asia.

Nevertheless, we still heard of the priest Gianni in the West in 1165, when the Byzantine emperor Manuel I Comnenus received a long, articulate and grotesque letter in which the sender described the immense territories of his kingdom, which, among other things, included: the three Indies, his palace built with gems and cemented in gold, the garden that was said to contain the fountain of youth and a magic mirror with which he could see everything that happened in his lands, and the table of his court, with 10,000 guests every day, whose footmen were made up of 7 kings, 62 dukes and countless counts, while the host of his subjects included, besides men, goblins, cyclops, minotaurs, dwarfs and giants.

The letter then went on to describe a cast of characters that would have made Hieronymus Bosch happy: cynocephali, monstrous creatures with the body of a man and the giant head of a dog; blemmi, headless beings with eyes and a mouth situated in the belly or chest; sciapodi, legendary individuals with only one leg and one giant foot that they would raise above their heads for shade.

The missive bore the signature of "John, Presbytery, thanks to the Omnipotence of God, king of kings and sovereign of sovereigns." The Byzantine emperor sent copies of the letter to Pope Alexander III and Frederick Barbarossa. The simplest hypothesis behind the delivery of this letter was, of course, that it was a joke, a message written by someone familiar with the mechanisms upon which the life of the courts and the concerns of the pope were based, perhaps someone in the inner circle around Frederick I or, as critics argue on the basis of the polished Latin and the numerous references to the Bible and Jewish beliefs, some learned ecclesiastic, an abbot of a monastery or an abbey. In fact, only someone in the clergy could have had that skill and linguistic imagination, hoping to push the pope to seek a powerful ally in the fight against the Muslims.

In fact, the pope, in order to get a clearer picture, decided to send his personal doctor as an ambassador to this fabulous sovereign; but when he arrived in Palestine, the doctor's tracks were lost and, with him, those of the priest Gianni. The text of the letter, however, became extremely popular, and hundreds of copies were made that reached all corners of the Christian world. Thus originated the myth that there existed, in some remote part of Central Asia, a kingdom of Nestorian Christians ruled by a king of immense power and, at the same time, with an odor of sanctity, a king who could play the role of fifth columnist behind the Islamic troops.

All of these stories enriched, with infinite variations, the legend of the priest Gianni that accompanied generations of travelers, pilgrims, chroniclers and rulers. For the pilgrims who went to the Holy Land, the reign of the priest Gianni was absolutely real, and they were only waiting for his troops to appear, fording the Tigris or the Euphrates, to save them from the assaults of the infidels. For a long time, however, there was no more news of this warrior priest, apart from Marco Polo's narrative, in which, however, he appeared under an entirely different identity.

Interest in this exotic kingdom and its legendary emperor-priest then slowly waned, until, in a real coup de théâtre, it had a revival around the middle of the fourteenth century, when the English traveller John Mandeville recounted that he had been to this fairy-tale kingdom during one of his trips. He left a manuscript of his memoirs to a doctor from Liège, Jean de Bourgogne, who published them, attracting enormous interest and wide circulation.

In those pages were all of the astonishing tales upon which medieval literature had already been amply nourished: golden palaces, shining stones, rivers of milk and honey, salamanders walking in fire and caverns full of dragons, as well as the entire great circus of monstrous beings that had been embellishing the facades of cathedrals for years. On his deathbed, however, the Belgian doctor confessed during extreme unction that the book was all an invention, a real fraud, conceived for the sole pleasure of teasing the stupid and crass credulity of the people.

Travelling in a Stubborn and Contrary Direction

How many journeys did Ulysses make? As many journeys as those of our Western civilization. The journey of the Odyssey is perhaps the most famous metaphor of the journey for knowledge, an endless wandering that coincides with the human quest to explain the order of the world, skilfully outlined in Canto XXVI of *Inferno*. It will be Virgil who invokes Ulysses and Diomedes, damned to Hell as fraudulent advisors, to respond to Dante's curiosity:

Let one of you declare,
where, lost, he went, to come upon his death.

Inferno, XXVI

And so begins the tale of that "last voyage." Ulysses.
confesses
[...] my long desire,
burning to understand how this world works,
and know of human vices, worth and valour.

Inferno, XXVI

He recounts a daring journey beyond the Pillars of Hercules when, after passing Seville and Ceuta, he had set himself a very precise goal: to gain "experience, behind the sun, of worlds where no man dwells."

Why does Odysseus want to go where no one has ever been? Well, we know at least this much: it was to "go in search of virtue and true knowledge." But why does he go in the direction of the southern hemisphere, below the equator, and not, for example, towards the last Thule, the legendary northern island, land of fire and ice, where the sun never set? Ulysses goes to the southern hemisphere to look for the "worlds where no man dwells" because he knows that God, the God of Christians, has forbidden any knowledge of it, just as he had forbidden Adam to pluck the apple of wisdom.

This idea of travelling to unknown lands had, in the past, aroused the curiosity of various writers, such as Lucian of Samosata, a skilled lecturer and expert Greek traveller who lived in the IInd century A.D., and who, in his *True Story*, had told of the fantastic voyage he would make towards the West, beyond the Pillars of Hercules, with narrative and linguistic details very similar to those of the "crazy flight" of Dante's Ulysses: the curiosity and desire for new things, the preparations, the persuasion of his companions, the high and wooded island, the storm. In the verses of the *Comedy*, one feels the silence of an endless ocean, the alternation of the moons, the anguish of the unknown that accompanies the night hours, the desperation on the faces of the sailors at the moment of the dramatic end of that senseless adventure, with the stern of the ship upwards and the prow downwards, until the sea closed in on those poor souls. Another great poet, Umberto Saba, gave word to Dante's Ulysses after he was swallowed by the waters: "Today my kingdom is that no man's land. The port lights up its lights for others; the untamed spirit still drives me out to sea, and the painful love of life."

A few years before Dante emphasized the tireless desire for knowledge of the hero Odysseus, another great traveller had gone in another stubborn and contrary direction, that of the Far East. It was Marco Polo. And Dante knew this, for Marco's adventures recounted in *Il Milione* were far better known than his unfinished *Commedia*. But Dante probably did not like *Il Milione*. By 1309, Florence had even produced a translation of Marco Polo's book that became a classic and was more widely reprinted than any other book. Dante was certainly not unaware of the Polo family's enterprise at the time, nor of the attempt by the Vivaldi brothers from Genoa to reach the Indies "per mare oceanum." And he certainly did not ignore what had been written by perhaps the greatest naturalist, physicist and astrologer of his time, Pietro d'Abano, a doctor from Paris and Padua, a friend of Marco's, who lived for a long time in Constantinople so as to learn Greek and Arabic, studying the original texts of Galen, Avicenna and Averroè.

Peter of Abano translated various Greek and Arabic scientific texts into Latin, enriching them with notes and additions. And he was famous as the author of *Conciliator Differentiarum, quæ inter Philosophos et Medicos Versantur*. Peter considered medicine to be the science of sciences, but he also paired it with the systematic study of astrology: it was essential to calculate the right time to administer cures and medicines, as well as having knowledge of the alchemy necessary for their preparation. In the *Conciliator*, he openly advocated the connection between the natural world and the stars, the use of magic, spells and medicine, and a conception of humanity and the whole of creation as a harmonious organism regulated by the constellations.

Dante knew these studies well and could not ignore the fact that, in discussing the 67th question or "difference" of his *Conciliator*—"whether life is possible under the equator"—the most illustrious naturalist of his time had affirmed, since 1303, the habitability of the southern hemisphere precisely on the basis of the accounts by Marco Polo, whom Pietro d'Abano described as "the greatest of all those who have travelled the world, investigating it diligently." From Marco, Pietro had received much news about the most characteristic phenomena of the equatorial regions he visited and had come to know the origin of the substances that he administered to his patients, such as camphor, aloe and "brasil," but he had also had confirmation of the existence of human beings and animals of considerable height in the regions of the torrid zone. No European before Marco had reached the borders of the great continental mass, a destination that had stimulated the curiosity and imagination of scholars and navigators for centuries, from Herodotus to Cosma Indicopleuste. Marco had seen with his own eyes what Dante's Ulysses had described hereafter:

Now every star around the alien pole
I saw by night. Our own star sank so low
it never rose above the ocean floor.

Inferno, XXVI

Marco had described how the island of Sumatra was located so far south of the North Star that the star had disappeared from the sky. The phenomenon had impressed him very much and he had returned to see the stars of the northern hemisphere only after he had rounded Cape Comorin, the southern tip of India.

If, for Dante, the southern stars heralded Paradise, for Marco, the reappearance of the North Star was an omen of his return home. Ultimately, Marco Polo's experience, corroborated by the recognition of a great scholar such as Pietro d'Abano, interfered with the scientific and metaphysical questions on the distribution of the habitable zones of the Earth and the configuration of the southern hemisphere and the antipodes, fundamental concepts of medieval cosmography and the Holy Scriptures. Like Dante's Ulysses, Marco had touched down not far from the hemisphere reserved by divine will for souls worthy of ascending to heavenly bliss.

By his silence, in our opinion well aimed, Dante condemned the narrow science of Marco and the overly broad science of Peter, two of the most famous Italians of his time. Like the sea that swallowed up Ulysses' ship, this silence lay over the memory of those who had crossed the limits of revealed knowledge, heralding the arrival of the age of humanity at the expense of the divine primacy of the knowledge of heaven and earth.

The Sect of Assassins

After leaving Palestine in April 1272, the Polos headed inland towards eastern Anatolia and Armenia, with the idea of crossing Persia, and then the land of the Tigris and the Euphrates, to finally reach the Strait of Ormuz, and, from there, embark for China. They had crossed the impassable paths of the bare Anatolian highlands on the backs of donkeys, and, having passed through a vast plain of Armenia, there appeared on the horizon, in all its majesty, the snowy massif of Mount Ararat, the mountain where Noah's ark had rested and where life had resumed after the universal deluge. All around, a light wind made the grass of the great valley sway and one could smell the strong scent of rosemary and thyme.

While wandering through the bazaars of the walled city of Ani, Marco had heard rumors of Georgia, a kingdom of grandiose mountains where eagles flew and the best hunting goshawks hunted. He diligently noted down everything he heard and saw: the dense fir forests, the impassable passes that had stopped even Alexander the Great, the lakes where there were fish to be caught, curiously enough, only in Lenten times.

Arriving in the districts of Iraq, he had witnessed with amazement the gushing of a black, oily, stinking liquid that flowed out among the endless plains of brushwood and stones and that burned with great ease, leaving immense pools as dark as pitch in the desert sand. "It came out in such abundance that a hundred ships at a time could be loaded with it; oil not usable as food but good for burning," he later wrote in *Il Milione*.

He would have liked to have visited Baghdad too, a city he had always heard wonders about, the beating heart of the Islamic world: all of the travellers he had spoken to had described its golden domes, its damasks and brocades embellished with finely embroidered animal designs. The city, they said, was inhabited by skilled magicians and equally cunning thieves who fled on flying carpets with sensual women. They had also told him that Baghdad was the city where the codes of the prophet Mohammed, necromancy, physics and astronomy were once studied in the madrasas, in short "the noblest and greatest city in that region."

However, thirteen years before Marco Polo's passage, Baghdad had been governed by a caliph famous for his greed and avarice, until the city was invaded by a horde of Mongol warriors under the command of Prince Hulagu. Having easily defeated the meagre army put in place by the caliph to guard the population, the Tartars entered the city and massacred almost all of its 70,000 inhabitants, sparing only those of the Christian community through the intercession of Hulagu's wife, a Nestorian Christian. When, at the end of the massacres, Hulagu learned of the exorbitant wealth of the caliph, he summoned him to ask why he had not used all of that gold to pay for an army to better defend himself and the population. The hesitation in his reply was sufficient to make Hulagu understand that it was avarice that had driven the caliph's arid heart; so, he had him locked up in a tower with all his gold, saying, "Caliph, eat of your treasure as much as you like, since you like it, and remember that you will eat nothing else, ever again." At the passage of the Polos after the Mongol conquest of the city, therefore, little remained to be seen of ancient Baghdad: there was only the faded memory of a great capital and a heap of rubble. They therefore headed straight for Tabriz.

The villages along the way were just a handful of houses gathered around a well, with a few almond trees and scattered trees with figs around them. From the stables rose the bleating of the lambs, from the chimney of the houses thin threads of smoke. In the shops of those villages, a thousand skilled hands worked the strong stone of the mountains and beat the iron, in a ritual that was renewed every day, always the same. In the distance, casting your gaze towards the mountains, you could see the outlines of the watchtowers and the bastions of some remote castle: these were lands that had always been used to wars and had never known peace.

Every day at dusk, Marco, together with his father Niccolò and his uncle Matteo, would seek shelter in the caravanserais scattered around, complexes equipped to house travellers, their beasts, and wagons with merchandise. They were all quadrangular in shape, their thick walls opening the way to a single entrance door wide enough for camels and horses to pass through; the larger ones had angular strongholds, a solid surrounding wall, underground cisterns where rainwater was collected, rooms for travellers and stables for the herds.

Those lands of Persia were rich in the most beautiful lapis lazuli in the world, balasci and turquoise, stones with green veins and dazzling colours, and in the courtyards of the caravanserais, one often witnessed animated discussions between jewellery merchants intent on sniffing out a bargain or avoiding a swindle. In the evening, it was a Babel of voices: the shouts in Arabic of the salt caravans were added to the thousands of languages of the Turkmen, the Armenians, the Persians, people who lived by buying and selling precious fabrics and carpets that arrived on the backs of mules and camels from the vast and distant territories of Asia. They were the offshoots of the long and articulated Silk Road that started from the remote corners of Catai.

All religions converged in those lands: together with the Nestorian Christians, there were the Armenian and Syrian Christians, the Persian followers of Zoroaster, and, of course, the Muslims, in the two sects of Sunnis and Shiites, divided by a hatred as dark and deep as Hell. At the origin of the schism in the Muslim world was the problem of Muhammad's succession, which arose when the prophet died in 632 without giving any indication as to the name of his successor. Apart from the religious and political legacy, there was also the enormous amount of money that flowed in from the various Arab tribes united under the banner of Islam.

While the majority of Muhammad's followers believed that his successor should be chosen by the community, an irreducible and rocky minority believed instead that only the descendants of the prophet had the right to take his place, and therefore it was Muhammad's cousin and son-in-law, Ali ibn Abi Talib, who took on the honour and the burden of the caliphate. These were called Shiat Ali—the party of Ali, the famous Shiites—while those who advocated the election of a member of the community to the role of caliph were called Sunnites: the latter, among other things, insisted that the choice fall on Abu Bakr, friend and father-in-law of the prophet. Since the exquisitely theological question had descended to purely family levels, it was inevitable that the disagreement would turn into anger, the anger into hatred, the hatred into a bloodbath.

This happened a few years after the death of the prophet, when the Shiites, led by Hussein, son of Ali, clashed with the army of the Sunni caliph Umayyad Yazid on the plain of Karbala, in Iraq: after days of furious fighting, Hussein's army was defeated and he was executed, together with his family. His body was tied to the back of a white camel that was left free to roam: after wandering for more than 2000 kms, the animal stopped near Balkh, an ancient Afghan city. There, the tortured body of Hussein was buried, finally finding peace. His martyrdom became, from that moment on, the central focus of the entire Shiite faith.

In addition to the schism between Sunnites and Shiites, it is necessary to underline that there was a further schism within that already much divided Arab world, a schism from which the sect of the Ishmaelites was born. Originally Shiites, the Ishmaelites, however, polemically broke away from that sect to take refuge in a series of castles and impregnable fortresses on the slopes of Mount Alamut, "the mountain of eagles," and from there unleash a war without quarter against those Seljuk Muslims who were guilty, in their eyes, of illegally holding power in the lands of the Middle East.

As was customary in that medieval world made up of stories and fables, this story also soon took on the features of a legend, that of the "Veglio della Montagna" (Better Man of the Mountain) and his sect of assassins. It was the classic story that was told in the evening, around the fire of the caravanserais, when people chewed pistachios and watched the flicker of the flame in the fireplace: it was the story that everyone wanted to hear again and again, because it spoke of worlds and things that everyone loved to fantasize about.

In the voices echoing around the bonfires, there were stories of fabulous secret gardens surrounded by golden walls, where, amidst perfumes and incense, rivers of milk and honey, and trees full of the sweetest fruits, "the most beautiful maidens in the world" played graceful melodies with reed whistles, dancing and singing. When the Veglio wanted to assassinate one of his enemies, the assassin was given hashish and it was promised that he would be taken to that enchanted garden as a prize: the hashashin, the "assassins," had acquired a sinister reputation throughout the East, and also in the West. The assassins were typically boys who, eager to return to the joys of Paradise that they had seen with their own eyes, were willing to do anything, even die, because death was simply the quickest way to return to seeing those graceful virgins dancing and singing in that fabulous garden.

In fact, news arrived from everywhere of carefully planned murders, carried out with ruthlessness and great lucidity: Conrad of Montferrat, heir to the throne of Jerusalem, was stabbed in a church; the same happened to a Christian king in Tripoli and numerous caliphs; it went better for Prince Edward, the future King Edward I of England, who was stabbed repeatedly in the alleys of Jerusalem, but managed to escape death. The assassins had quickly earned the fear of the caliphs, princes and rulers, who, to avoid being disembowelled by these killers, did not hesitate to go to the Veglio to pay him homage and money in order to earn his protection.

This climate of terror ended with the arrival of the Tartars, who were terribly annoyed by all of that arrogance. Hulagu, the same man who had turned Baghdad over to arms, pursued the systematic annihilation of the sect: he conquered all the Ishmaelite castles one by one until the final capitulation of the last fortress on the slopes of Mount Alamut, after a siege that lasted three years. And, as in Baghdad, Hulagu mercilessly killed the Veglio and all his followers: "Together with the Best, the castle and the garden of Paradise were flattened. From then on, there was no Veglio and there were no more murderers: and with him ended forever the evil rule imposed by the Vigils of the Mountain."

Southern Cross

As we have mentioned, Dante began writing the *Purgatorio* between 1308 and 1309, and published it between the end of 1315 and the first half of 1316. The temporal continuity of the writing between *Inferno* and *Purgatorio* better highlights the political and ideological discontinuity of this Cantica with respect to *Inferno*, in which the Empire was spoken of in a strictly Guelph sense: if, in *Inferno*, Dante had censored himself to try to obtain amnesty, in *Purgatorio*, he returned instead to an evident pro-imperial vision.

In all probability, the first half of *Purgatory* had been written in haste by the summer of 1310, including Canto XVI, in which Marco Lombardo appeared, with his political theory of the two Suns (emperor and pope). After the death of Corso Donati, however, Dante was now sure that he would not be able to return to Florence. The *Purgatory* therefore included a staged political palinody. In Canto III, Manfredi, the natural son of Frederick II, was described with sympathy for and emotional participation in the imperial cause. After the disappointment of Guelphism, Dante himself reappeared, and, in the *De vulgari eloquentia* of 1304, likewise mentioned "Frederick emperor and his worthy son Manfred."

Purgatory opens with a rediscovered blue sky of the "Soft hues of sapphire from the orient" when:

The lovely planet, strengthening to our love,
lit up with laughter all the orient sky,
veiling her escort, Pisces, in bright light.

Purgatorio, I

The darkening of the constellation of Pisces by Venus indicates, with astronomical exactness, that we are at dawn on the fourth day of the voyage, March 28, 1300. Needless to say, this and the subsequent astronomical reference integrate the religious message of the poem on the basis of an undoubted competence. Competence that becomes rare in the reference to the stars of the Southern Cross:

turned now to the right. I set my mind
up on the southern pole, and saw four stars
that none – save Eve and Adam – ever saw.

Purgatorio, I

There has been much discussion about the "four stars" and the first people. Allegorically, the four stars are the cardinal virtues (prudence, justice, fortitude and temperance), which are presented to Dante in Canto XXXI as maidens singing: "We, here, are nymphs and, in the heavens, stars." They appear in the morning because they represent the active life. As for the first people, it seems clear to us that the reference is to Adam and Eve, if we think that original sin took place in the Earthly Paradise, and is therefore in that latitude.

The Southern Cross was little known in the Middle Ages and was seen directly only by those who had crossed the equator, such as Marco Polo. In the *Conciliator*, the afore-mentioned Pietro d'Abano reported having spoken with Marco Polo about what the Venetian had observed in the sky during his travels, in particular, a star "shaped like a sack" (ut sacco) with a large tail (magna habet caudam), of which he had kept a drawing.

Some clarification of the sense that d'Abano gave to the testimony of Marco Polo may come from examination of the testimony of another traveller whose account Pietro went into in more detail. D'Abano had received a letter, arriving from the Coromandel Peninsula, north-west of the North Island of New Zealand, from a certain frater Iohannes Cordellarius, who can be identified with the Franciscan Giovanni da Montecorvino. The letter, rendered in the Tuscan vernacular by one of the addressees, Friar Menentillo of Spoleto, must also have circulated in Franciscan circles in Padua.

Friar Giovanni believed that, from a high place, it would be possible to see the star to the south that was opposite to the polar star. D'Abano used this letter extensively, because he considered it much more complete and specific on the astronomical level than Marco Polo's testimonies. Friar Giovanni's letter was, in fact, full of measurements regarding the position of the stars, which were scarce and generic in *Il Milione*. D'Abano intended to refer explicitly to what the cartographers of the time referred to as "the other tramontana," the Southern Cross. It can be assumed that the Paduan scholar had come into possession of the letter in the days when he visited Marco Polo, probably during the summer of 1303.

For these and other researches, Pietro d'Abano was persecuted as a heretic, both during his life and even after his death. The Tribunal of the Inquisition issued 53 charges against the *Conciliator's* theses, including necromancy and Averroism, but also accused him of questioning the miracles of the saints and the existence of demons. A year after his death, in 1315, the Tribunal decreed that the remains of the scholar should be burned at the stake: when it came to certain issues in the fourteenth century, people were not very subtle, and not even death could extend its protective shield.

Returning to Dante's voyage, the first triplet of Canto I of the *Purgatorio* is allegorical and makes use of seafaring language:

To race now over better waves, my ship
of mind – alive again – hoists sail, and leaves
behind its little keel the gulf that proved so cruel.

Purgatorio, I

It is the morning of Easter Sunday and Dante hopes that "dead poetry will rise again," an indication of the new state of mind of those who come out of the infernal cavern and begin a journey of salvation. Even the sun, which represents divine goodness and grace, is preparing to rise again, indicating the path of purification, in opposition to what happened at the entrance to Hell, in Canto II, when the sun was setting.

The planet Venus represents the love between souls and between them and God. The sins punished here are all sins of love, sins that can go astray "when wrongly aimed" (pride or self-love, envy or love of evil, others, wrath or love of revenge), for "lack [of] vigour" (sloth or laziness, that is, a deficit of love of the good) or for "too much vigour" (avarice and prodigality, i.e., an overabundant love of wealth, or gluttony, i.e., an overabundant love of physical sensation, lust or sex), according to what Dante discusses in Canto XVII.

Dante and the early astronomers

Mary Acworth Evershed, also known as Mary Orr—the maiden name by which she signed her book *Dante and the Early Astronomers*, had two great passions, astronomy and the *Divine Comedy*. It can be said, throughout all her life, she sought out poetry, both in the verses of the Poet and in the darkness of the vault of heaven. Her story is a small gem in the history of astronomy, one that also allows us to take a look at that eclectic world of the English colonial age in the Victorian and Edwardian periods. Mary Orr had been born in the coastal town of Plymouth, England, in 1867, and had come up with a curiosity that was out of the ordinary. The *Comedy* had begun to circulate in England following its translation in 1818 and, reading it, Mary had been quite struck by that singular poem and by the dramatic story of the man who had lost everything, home, family, possessions, and had been forced to wander in exile until the end of his days. In her twenties, Mary made a long trip to Italy that changed her life forever: she visited the most important of Dante's places, Florence and Ravenna, and began to examine, with great care, all of the astronomical references in the *Comedy*, not only to catch errors, but also to find the poetic prologue to her future astronomical discoveries. We know that the astronomical allusions in Dante's poem are numerous and also sufficiently detailed: T.S. Eliot had dismissed them as irrelevant for appreciating the poem and "almost unintelligible." Mary Orr saw it differently: that poem was the work of someone who had spent hours observing the sky, someone who had a broad and deep understanding of the astronomical science of the thirteenth and fourteenth centuries.

She discovered, in fact, that Dante had not made any flagrant errors, and that those occasional little mistakes of his were a reflection of the faults present in the almanacs available at the time: the man had taken the trouble to calculate the exact time of the revolution of the Sun around the Zodiac, had never made the mistake of speaking of a crescent moon rising in the east at sunset, and had spoken of the punctuated disk of the Sun three centuries before sunspots were officially discovered by Galileo Galilei! Mary Orr was fascinated by Dante's faithful descriptions of the vault of heaven, based on what he had learned from Aristotle, Ptolemy and Euclid, and understood that the Poet had the same impulses as a scientific researcher, the same restless mind, the same habits of making precise and methodical observations. At the beginning of the twentieth century, women wishing to become professional astronomers faced many obstacles: Mary Orr joined the British Astronomical Association as an amateur astronomer and met John Evershed during an expedition to observe a solar eclipse. Like Mary, John had been drawn to astronomy while still young and was so highly regarded for the expertise of his observations that he was invited to direct the Observatory of Kodaikanal, situated high in the Palani hills of southern India. It was on the occasion of that appointment that John asked to Mary to marry him, after five years of friendship spent watching the sky together.

They left for India in 1906, and, within that lonely outpost of the British Raj, the two became a formidable scientific team, all of their efforts directed at unlocking the secrets of the cosmos, particularly those of the Sun. In the evenings, Mary sat on the hillside gazing at the Moon and the constellations of the southern hemisphere. The forests and lakes with waterfalls, where

> plum and plane trees grew alongside rare kurinji flowers, reminded her of the geography described in sections of the *Comedy* and were a source of great inspiration to her. She had to wonder: Was Dante's astronomy really incomprehensible? She thus began to work diligently on her astronomical project on Dante, which was eventually published in 1913 under the title *Dante and the Early Astronomers*. The book was reviewed in passing in a couple of astronomical journals during the First World War, but it was not until the 1940s that it was rediscovered by mystery novelist Dorothy Sayers while she was working on a translation of Dante for Penguin Classics. The Eversheds returned to England in 1923, and Mary went on to direct the historical section of the British Astronomical Society, publishing a volume entitled *Who's Who in the Moon*. Like her work on Dante, it was a fusion of astronomy and philology: an account of the origins of crater names, up until at least the late 1930s. Again, an admirable synthesis of the two cultures, the scientific and the humanistic, united by the same majestic and refined link present in the *Comedy*.

But there is, in Dante, something more than an observation, albeit a precious one, of the vault of heaven. There is a reversal that is also a change in time and space: "I turned to the right," he writes, while, in Hell, he had always walked to the left. In Purgatory, time flows inversely, because it leads towards eternal life. And the substance of the reality that presents itself before Dante's eyes also changes: the four earthly elements (earth, water, air, fire) are replaced by that incorruptible and eternal element that is the ether. It is still a world of visible and corporeal things, but they, inasmuch as they are made up of ether, are unalterable and eternal. And this is well demonstrated, in Canto II, by the soul of the musician Casella, whom Dante tries in vain three times to embrace.

Purgatory is also a world of light, as opposed to the darkness of Hell: the natural light of reason, personified in Virgil, will be replaced by reason illuminated by divine truth, represented by Statius the poet. A new journey begins that marks a detachment from the leaden and overly earthly world of Hell in order to undertake a purifying asceticism towards the vision of the divine.

In the Footsteps of Alexander the Great

The lands of the East were overflowing with history, as Marco Polo personally realised as he went further and further into the regions south of the Caucasus. The dividing line between the West and the mysterious heart of Asia began with the mountains that sloped down towards the Caspian Sea, north of what was called the "land of the wolves." On these mountains, there was a fortified pass, very narrow and impassable, known as the Iron Gates, with one fortress at the entrance and another one at the exit of the valley, as well as manors built on the first fortifications put up by Cosroe in the sixth century to protect the northern border of the Persian Empire.

Those Iron Gates, however, were universally associated with the legendary deeds of Alexander the Great: it was Alexander himself who had created his own myth thanks to the flood of stories and tales written by the host of historians and chroniclers who accompanied his military campaigns. In just twelve years, he had conquered an immense territory stretching from Asia Minor to Egypt, and even as far as Afghanistan and northern India, a heroic feat that only confirmed the rumours that he was a descendant of Heracles and Achilles, or even real gods. The fame of Alexander the Macedonian had remained alive in those distant lands of Asia thanks to the thousand storytellers in the bazaars, and was even consecrated by the biblical tales found in the First Book of Maccabees and some passages of the Koran. Marco Polo recalled that, when he went to St. Mark's Basilica in Venice, he was enraptured when he looked up and gazed at the bas-relief on the northern facade of the church in which the great Alexander was lifted into the air by two griffins.

According to legend, the Macedonian conqueror had erected the Iron Gates to bar the way for Gog and Magog, the horde of the devil charged with destroying the world and bringing chaos upon it. If, in the Apocalypse of John, we read that these savage and bloodthirsty monsters coming from the darkness of the Asian deserts were in charge of the destruction of Israel, but would be defeated by Christ in the definitive battle against Satan, the Koran instead recounted that these giants had been confronted and defeated by the Macedonian, who had imprisoned them behind a copper wall beyond the Iron Gates. The rest of the Koranic tale, however, was anything but reassuring, with a final sequence bearing all of the pressing rhythm of a horror movie: it was written, in fact, that, one day, these monsters would escape and would rush to exterminate every human being on Earth; once this carnage was over, they would make the mistake of turning their anger towards the sky, launching poisoned arrows and darts upwards, an act that would unleash God's wrath and lead them to a miserable end. The Almighty would hurl worms at them, which, having penetrated into the noses and ears of those demonic beings, would kill them by devouring them from the inside.

The Iron Gates were not the only legend about Alexander the Great: there were also the mythical palaces of Persia and India, the fountain of youth, the stories of the strange and monstrous peoples he met in the most remote corners of the planet, the magical forests and the fantastic animals that populated them. And, travelling in those Eastern lands, Marco heard from the inhabitants of continental and tropical Asia all of the same stories about Alexander the Great that he himself had heard in the Venetian squares from passing bards.

While crossing the province of Tunocain, rich in cities and castles, Marco had also come across the Dry Tree, another of the great Alexandrian legends. This tree stood, dry and lonely, in the middle of an endless and desolate plain bordering Persia on the north side:

> *It is immense and mighty, with green leaves on one side and white on the other. It produces curls similar to chestnut curls, but they are all hollow inside. It is of strong, yellowish wood like boxwood. All around for a radius of a hundred miles there are no other trees except in one direction where grow trees at a distance of ten miles. And this is the place where, according to those of the country, the battle between Alexander and Darius took place.*

Marco notes everything in a prose made up of short, precise, almost didactic sentences. This Dry Tree of which he speaks was a plane tree, considered sacred by the inhabitants of the plain that opened up at the foot of the Elbruz mountain range. Ever since the beginning of time, epic battles, decisive for the fate of the world, had always been fought at the foot of this tree, starting with the battle between Alexander and Darius III of Persia. For Christians, this tree had been placed near one of the rivers of the Earthly Paradise as a symbol of the knowledge of good and evil, but it had dried up as soon as Adam and Eve had committed the original sin, and all that remained around it were sparse pools of brackish water hidden among the dunes.

In the arid lands of Persia, the Polos rode for miles and miles along mule tracks swept by wind and sand, being guided by the sun: at Sapurgan, they found that the locals, to quench their thirst and feed themselves, cut melons into thin slices and let them dry, which made them sweeter than honey. The echo of Alexander followed them wherever they went: in Balkh, the inhabitants boasted that they lived in the oldest city in the world, the one where Alexander had wanted to marry Statira, the daughter of Darius, the king of kings.

The province of Badakhshan, perched on the Afghan mountains, also boasted of being a kingdom with direct ties to the Macedonian: the men of that district boasted of having traded horses directly descended from Bucephalus, the superb animal of the best Thessalian breed tamed by Alexander as a young man who then became his faithful and inseparable companion in battle. The horses of the region, in fact, all had a mark on their foreheads, almost a mark of royalty that came from the recesses of history: Bucephalus had a black coat and a white star on his forehead, and his eyes were of two different colours, one bright blue, the other as black as coal. The legend told by the people of Badakhshan was that, in the past, the horses were bred only by an uncle of the king, who took care of them as if each was the same legendary Bucephalus. The sovereign's eagerness to be a protagonist, however, had led to a tragic epilogue of that noble and millenary art of horse-riding: the king wanted to have all the steeds in his possession, but his uncle had opposed him, issuing a vulgar refusal, and for this, he was sent to the gallows. The widow, in revenge, had all the horses killed, thus putting an end to the glorious lineage of Bucephalus.

Along the caravan routes of Persia, in the oases of the deserts and in the bazaars of the cities, other exotic stories echoed in addition to the deeds of Alexander, such as the one involving the Magi who had gone to Bethlehem to pay homage to "the birth of a new prophet." The worship of these three wise men was very much alive among the people of the Persian highlands, who fervently followed the creed animist of Zoroaster.

Marco picked up the story from the inhabitants of Cala Aperistan, a name that meant "castle of the fire worshippers": the three kings had decided to bring gold, frankincense and myrrh as an offering, so that they could recognize whether that prophet was God, king or sage: "…they thought: if he takes gold he is a king, if he takes frankincense he is a God, if he takes myrrh he is a sage." When they arrived in Bethlehem, the youngest of the three went to visit the newborn child, but he was astonished, because he discovered that the child looked surprisingly like himself and seemed to have the same age and appearance. The same scene was repeated in the individual visits of the other two. So, they decided to go all three together into the stable where the child was lying, and this time he appeared to them for what he was, an infant only a couple of weeks old: "They worshipped him and offered him gold, frankincense and myrrh, and the child took all three offerings, and gave them a closed casket. And the three kings departed." After several days of travel, their curiosity became unquenchable. They finally opened the box, only to find a stone in it!

For Marco Polo, the interpretation of the story was clear: since the child had taken all three gifts, he could only be God, king and sage at the same time; on the other hand, the child, having seen that faith had been born in the three Magi, had given them a stone so that they would remain firm in what they had come to believe. However, not truly understanding the meaning of that gift, they took the stone and threw it into a well. From heaven, then:

A burning fire came down and went straight down into the well. When they saw that miracle, the three kings were astonished and regretted having thrown away the stone; they had understood that it was in reality a great and admirable sign. So they took that fire and brought it back to their country to keep it in a beautiful and rich church where it has burned ever since, adored as a God.

So, the story of the Magi, in the words of Marco Polo, becomes, as if by magic, a cross-section of the history of the religions of medieval Asia, an unorthodox version compared to the very popular one propagated by the Gospels. The stone thrown by the Magi into the well was, according to Marco, the origin of the cult of fire in Persia and of the existence of those very ancient Zoroastrian communities that were later swept away by the Islamic invasion, and of which only traces remained in the Parsis in northern India and in very few other Persian villages. Actually, the myth of fire linked to Zoroaster is much more ancient: fire, purifier of the world's evils and symbol of the divine light itself, prerogative of the Creator's chosen ones, had been the perennial obsession of Zoroaster, an Iranian prophet who lived around 600 B.C., that is, many hundreds of years before the legend of the Magi.

In the course of his journey along the transversal route leading to the heart of Persia, Marco passed by the magnificent city of Yazd, of which he maintained a vivid memory: in addition to the sophisticated system made up of articulated tunnels that conveyed water from the mountains into the city, he noticed that the urban landscape was broken up by the sky-blue tiled domes and the Wind Towers; the latter were tall buildings with which, thanks to a system of cisterns and the circulation of humid air, the houses were ventilated, thus pleasantly relieving the premises of the torrid air that oppressed them. The most disturbing profiles, however, were those of the Towers of Silence of the fire worshippers: on the roof of these imposing buildings, the dead were laid in open circular pits and left to be eaten by birds of prey, in service of the eternal dance of life and the return of the remains to nature, according to the tradition of the celestial burial preached by Zoroaster.

The Fire Purgatory

The writing of the *Purgatorio* slows down after the summer of 1310, because, between 1311 and 1313, Dante was involved in the unfortunate adventure of Arrigo VII. A now disconsolate consideration of the Italian condition corresponds to the lack of prospects that characterised the period between the death of Corso Donati and the descent of Arrigo VII. Consequently, in the last third of the *Purgatorio*, Dante takes a path that moves away from contemporary historical events and turns to the Christian sense of his purifying ascent, from an increasingly lyrical perspective. Dante's prophecy explodes in the last Cantos of the *Purgatorio*, when the Poet arrives at the summit of the mountain and enters the Earthly Paradise. Purgatory was born, poetically and culturally, as a result of Dante Alighieri taking up themes that had been mentioned only vaguely by the Holy Fathers of the Church up until then. The purgatorial tradition envisaged a purifying fire and, with Augustine, the practice of suffrage for the souls of the dead began. Traces of collective commemorations of the dead are reported as early as the seventh century in Seville and the ninth century in Fulda, Germany. Between 1000 and 1009, Saint Odilon, abbot of Cluny, instituted the feast of the commemoration of the dead, which spread rapidly in France and the Nordic countries, arriving in Italy in the thirteenth century and in Rome at the beginning of the fourteenth century.

The monk Jotsuald, in his biography of the holy abbot of Cluny, recounts a legend that would also be repeated by Jacopo da Varazze in his *Legenda aurea*, a very popular work at the time of Dante: a monk returning from Jerusalem, caught in a storm off the coast of Sicily and stranded on a rocky islet, found a hermit who told him that, within a nearby burning volcano, the purging souls were punished, and that these souls invoked prayers from the living for the remission or shortening of their punishment. The feast of the commemoration of the dead, the day after the feast of All Saints, is said to derive from this event.

One of the churches dedicated to the souls in Purgatory was built in the port of Lipari, in memory of the legend that placed Purgatory in the Aeolian Islands. The role of Sicily as the penal seat of the afterlife is confirmed by various legends; in contrast to Ireland, where it was believed that Purgatory could be found in St. Patrick's well, the interpretation prevailed in Sicily that Hell could be found there, in continuity with the pagan tradition that had placed the infernal forge of the god Vulcan inside Mount Etna.

In the particular religious climate of the late Middle Ages, the doctrine of Purgatory developed and was defined: according to Saint Augustine, the fire in the afterlife would be more terrible than anything that humans could suffer on earth. Dante was a boy when Purgatory was discussed at the Second Council of Lyons and the doctrine was accepted by the Latin Emperor of the East, Michael Paleologus. It was at that point that the term "purgatory" became a noun, indicating a place and a condition, after having previously been attributed to "fire," to indicate the purifying fire of the afterlife. Dante, for his Purgatory, draws on the theological debate and some of the legends of his time: in ancient commentaries of the Bible, the place of purification was located on a very high mountain, an idea that he takes up, but he is the first to give an organic arrangement to the realm of purification, with the division of the mountain into Antipurgatory, Purgatory proper and the Earthly Paradise.

Dante's purification, presented in Canto I of *Purgatory*, is willed by God, through the intercession of Beatrice. Virgil tells him clearly:

Yet if in Heaven there is, as you declare,
a lady who commands and moves your deeds,
fine words aren't needed.

Purgatorio, I

Virgil must first purify Dante of the filth of Hell, following the instructions of Marcus Porcius Cato, who oversees his entry into Purgatory. Dante's rite of purification requires "un giunco schietto e che li lavi 'l viso, sì ch'ogne sucidume quindi stinghe"; the reed symbolizes the humility of the sinner, who must prepare himself for the new way; the plant torn up by Virgil is immediately reborn in the same place and manner. Virgil cleans Dante by placing his hands on the grass and collecting drops of dew with which he washes the Poet's face, preparing him for the ascending journey. The rite ends on that deserted beach, in front of that sea that has never been navigated by a human being, however experienced. The reference, by no means casual, is to Ulysses, the one who had placed superb trust in his human abilities, but who nevertheless became lost.

In Purgatory, there are sunrises and sunsets; it is the only region in Dante's work that allows for chiaroscuro. Dawn is always an invitation to the journey: purifying and ascending, it produces a feeling of trust, hope and expectation. Sunset urges meditation, while night is absent. We have witnessed the first dawn of purification in Canto I: the last dawn appears between Cantos XXVII and XXVIII, that is, when Virgil says goodbye and leaves Dante, having fulfilled his task. Having thus reached the threshold of Paradise on Earth, Dante finds himself before an extraordinary idyllic aura, illuminated by the first light of the sun:

A gentle breeze, unchanging in itself,
struck on my forehead, yet with no more force
than would the smoothest of our changing winds.

Purgatorio, XXVIII

It is Dante's truest dawn, that of one who has finally reached the salvation of the soul in divine grace.

The Miracle of the Mountain

If *Il Milione* were only a dry list of cities and places visited by our Venetian, it would not be that extraordinary a work, admired, read and reread, copied in numerous illuminated editions and, ultimately, representative of the transparency and mystery of "a spring water in a crystal bottle" that we cherish.

One of the book's virtues is that it mixes, with disconcerting naturalness, precise geographical annotations with stories shrouded in legend and the colours of fantasy. Like the utopian world maps of the early Middle Ages, where, to the multitude of mountains, rivers and cities, were added special places inhabited by magicians, soothsayers and mythological characters, so, too, Marco willingly gave in to the pleasure of storytelling, especially if it could cast the followers of Mohammed in a bad light.

If it is true, in fact, that he generally showed an unusual tolerance for all religious beliefs and customs of various peoples, however strange, it is also true that he could not hide his dislike and distrust for Muslims. Thus, the Turcomannians, who adored Mohammed, are described as "primitive people who speak a rough language," the Mohammedan Kurds as "warlike and ferocious people who often plunder merchants." So, if he could ridicule them by telling the stories that he had learned from the Christians who populated the Middle East, so much the better. On the other hand, Persia was a real gold mine of stories, given the perennial integration of its population and the succession of wars for the conquest of those lands: just think of the fact that *The Arabian Nights* were born there.

One of the most famous stories reported in *Il Milione* is "the story of the miracle of the mountain," which took place in 1225, between Baghdad and Mosul, and had as its protagonists a powerful caliph and a humble shoemaker. The Caliph of Baghdad in question hated Christians, and thought day and night about how to force them to become Saracens, or, if they refused, how to put them to death, taking counsel with the other leaders of his religion, "for it is true that the Saracens of the whole world are bitter enemies of the Christians of the whole world." Macerated in this obsession of his, he studied his enemies: reading the Gospels with great attention, he stumbled upon a very interesting verse, so interesting that, as soon as he read it, his face was crossed by a sinister grin and his eyes shone with a fiercely malignant light. In the Gospel of Matthew, there was a curious statement made by Lord to the apostles: "Verily I say unto you, If ye have faith as a mustard seed, ye shall say unto this mountain, Move from hence to thither, and it shall move, and nothing shall be impossible unto you."

He had finally found the way to convert them or kill them all! He summoned all the Nestorian and Jacobite Christians to his palace, showed them the Gospel, and questioned them as to whether that book spoke the truth. When they answered in the affirmative, he repeated the verse aloud and asked if it was true that, by prayer alone, a man of Christian faith was able to bring the mountains closer. The answer he received was again in the affirmative:

Here then is my proposal—resumed the caliph—you Christians are many and certainly among you there is a man who has some faith. Well, this is what I say to you: either you make those mountains move [pointing to the mountains that stood out in the distance] or you will all die a miserable death. For if you do not make them move, it will mean that you do not have even a grain of faith. Then either you will convert to the good religion which Mohammed our prophet has given us, and, having the true faith, you will be able to save yourselves, or you will die. I give you ten days, and if after that time you have not brought the mountains nearer, I will have you killed.

And this is where the shoemaker came in. Eight days and eight nights after that terrible conversation with the caliph, the archangel Gabriel appeared in a dream to one of the bishops: the angel said that the solution was an old one-eyed cobbler. This cobbler was a most upright Christian who went to church every day and shared his bread with the poor. He had become blind by his own hand, because he had read in the Gospel that, if the eye was the cause of a sin, it would be necessary to remove it in order not to incur the evil again. One day, a beautiful woman came to his shop and asked him for a pair of shoes: he asked to see her foot and leg so that he could choose the right size, but he could not imagine that the sight of her would affect him so strongly! Feeling that his eyes were sorely tempting him by looking at the woman's flesh, he abruptly sent the woman out of the shop and, taking a chisel, "hit his eye so that it burst in the head; and from that eye he could no longer see." This was the holy and virtuous man indicated by the archangel Gabriel, and on whom all the hopes of the Christians were placed. They went to him, begging and entreating him to pray to the Lord. Their begging was so adamant that he agreed.

The tenth day arrived, the crucial one. All of the Christians woke up in good spirits and gathered around the bishop for holy mass. Then, they went in procession, with one carrying a large cross at the head, and reached the plateau outside the city that opened up at the foot of the mountains. There, the caliph was waiting for them, together with a great number of Saracens armed with spears and scimitars:

> *The shoemaker knelt down and stretching out his arms to heaven prayed fervently to the Savior that he would make the mountain move so that all those people gathered there would not die a cruel death. He finished his prayer and the mountain moved, beginning to shake and advance.*

In the golden sky of that day, a miracle occurred: the thunder of the advancing mountain left the caliph appalled and immediately converted many Saracens to Christianity. The caliph also converted secretly: "Only when he died did they find the cross hanging from his neck; for this reason they did not bury him in the tomb of the other caliphs but put him in another place." The mysteries and miracles of the East.

Dreams, Prayers and Numbers

In *Purgatory*, Dante fills the gaps between sunset and sunrise with the narration of his dreams. In Canto VIII, he recounts his first dream after falling asleep on the flowery meadow, beginning with the famous triplet:

> *It was, by now, the hour that turns to home*
> *the longing thoughts of seamen, melting hearts*
> *the day they've said goodbye to dearest friends,*
>
> Purgatorio, VIII

At that time, in fact, dreams experienced at dawn were believed to be true. Dante dreams of "an eagle in the sky. Its plumes were gold": the eagle seizes him on Mount Ida, carrying him up high. Virgil explains that it was Saint Lucia, who wanted to lead him to the door of Purgatory to enable him to save himself. Dreaming of crossing the sphere of fire with the eagle, Dante wakes up:

> *And there the eagle and I, it seemed, both blazed.*
> *And this imagined fire so scorched and seared*
> *that, yielding, dreaming sleep just had to break.*
>
> Purgatorio, IX

This, perhaps, because he has, "in reality," felt the warmth of the sun, already high on the horizon. This first dream anticipates two others, which Dante will have over the other nights spent in Purgatory, in Cantos XIX and XXVII.

In Canto XIX, before passing through the fifth circle where the penitents for avarice are found, again at dawn, Dante has a dream of a deformed woman, warlike, pale, slovenly, stammering; as the Poet looks at her, however, she transforms, straightening up, taking on colour and regaining the power of speech. She tells him, among other things, that, in the past, she had also attracted Ulysses with her singing. But here, suddenly, appears a holy woman, who assaults her, tears her clothes and uncovers her belly, unleashing a stench so fetid that Dante wakes up. The "stammering crone" is the lure of earthly goods that induce incontinence; the holy woman is the philosophy that unmasks the deceptions of the passions.

The third dream is found in Canto XXVII, when Dante is about to enter the Earthly Paradise. Here, as we have already mentioned, he dreams of a young and beautiful woman wandering in a plain, singing and picking flowers: this woman, Lia, is the biblical sister of Rachel, whom exegesis interpreted as an allegory of the active life, while her sister was a symbol of the contemplative life. Lia's attitude anticipates that of Matelda, whom Dante will meet in the next Canto. The active life is indispensable for the attainment of the cardinal virtues, which lead to earthly happiness, symbolized by Eden. The dream, therefore, prefigures Dante's entrance into the terrestrial Paradise.

Alongside the premonitory and symbolic dreams, in Purgatory, the realm of penitence, there are numerous hymns and prayers that the souls sing or recite, always with specific meanings, with no less than 29 of them being sung according to the various occasions: of these psalms, the Poet refers only to the first words, because, since they were sung by everyone in church, the first verse was enough to put them in mind. For example, in Canto II, the souls coming from the mouth of the Tiber, symbol of the Church of Rome as the sole procurer of eternal salvation, as soon as they disembark from the "boat so swift, so quick, so light, so elegant," intone the psalm "In exitu Isräel de Aegypto," which recalls the miraculous exit of the Hebrews from Egypt and from servitude to the Pharaoh, a hymn once sung during the transportation of the dead to the cemetery. The Hebrews represent God's faithful, Egypt represents the state of subjection to sin and the Pharaoh symbolises Satan. The hymn is therefore a symbol of liberation from sin for the souls on their way to Purgatory.

Even more important than song is the theatre, in the medieval form of the sacred representation, clearly present in the last Cantos of *Purgatory*, with the classic partition into a prologue, three acts and an epilogue. From the theatre, this representation has a proscenium, music, dance, choruses and a particular way of proceeding. To the greenery and flowers of nature are added lights, sounds, songs, saints, angels, monsters, costumes and movements, in a symbolic context of exceptional deployment, which is also reflected in processions and liturgical rites, in which every word, color and gesture is easily understood by all of the people, accustomed to "reading" the paintings, frescoes and mosaics of churches (the so-called "Bible of the poor"). Let us look at an example of such a purgatorial representation.

In Canto XXIX, in the song "Beati quorum tecta sunt peccata!", a mystical procession begins, for which Dante seems to have drawn on Saint Jerome's *Epistula ad Paulinum*, which reviews the books of the Bible. The overall design of the procession appears as a cross, the centre of individual and world history. Eve, "a woman, one alone, formed only now," symbolizes the unveiling of the knowledge of good and evil, the object of divine prohibition. Seven lit golden candlesticks then appear before Dante's eyes, symbolizing the gifts of the Holy Spirit: wisdom, intellect, counsel, fortitude, knowledge, piety and the fear of God. They are opposed to the seven deadly sins and begin the procession.

Dante also sees a trail in the sky composed of the seven colors of the rainbow, a symbol of brotherhood and peace, representing the effects of the seven gifts, that is, the ardor and light of the Holy Spirit, or the seven sacraments. And who knows whether the Poet might not just have been imagining a painting or a miniature? So, there "were elders—twenty-four—who walked in pairs, and each of them was crowned with fleurs-de-lys," symbols of the twenty-four books of the Old Testament, representing twelve patriarchs and twelve apostles, and crowned with lilies, symbol of the purity of biblical doctrine or faith in the coming of Christ. The "four animals, and each of these was crowned with boughs of green," are monsters, representing the four Gospels, the green fronds with which they are crowned symbolizing the everlasting youth of the evangelical doctrine, their six wings the triple interpretation of the Scriptures beyond the literal one (allegorical, moral and anagogical) and the feathers full of eyes the acute intellectual sight of the evangelists.

Next is "a two-wheeled chariot in triumphal state," representing the Church. "A gryphon drew it, harnessed at the neck": here is another monster, with the head and wings of an eagle and the body of a lion, representing Jesus Christ, with his two natures, divine, "The bird-limbs of that form were all of gold," and human, "the others white, commingling with bright red."

On the right of the chariot, "three ladies came, all dancing in a ring," symbolising the three theological virtues: the red one is charity, which governs the dance of the trio, and therefore also the quality and quantity of faith, represented by the white one, and of hope, the green one. The four women on the left of the chariot, "all purple-clothed," symbolize the four cardinal virtues and proceed being lead by Prudence, who has three eyes on her head, thus being able to scrutinize the past, present and future simultaneously. The "two elders, differing in their garb, but equal in demeanour, grave and firm" are Saint Luke, author of the Acts of the Apostles, and Saint Paul, author of the Epistles. The "four, each one with humble looks," are Saints Peter, James, John and Judas, and the "old man all alone [...] face alert" who follows them is Saint John the Evangelist, absorbed in the visions of the Apocalypse.

The thunder, a sign of divine power, which concludes the Canto, manifests an imminent prodigy, which will occur in the next Canto, in which Beatrice appears to Dante. The sung invocation, "Veni, sponsa, de Libano!", taken from the *Canticle of Canticles*, symbolizes, in the bride, the Church or Beatrice herself, the emblem of revealed truth. The apparition of Beatrice, who takes the place of Virgil as Dante's guide, represents the climax of the sacred representation:

So now, beyond a drifting cloud of flowers
(which rose up, arching, from the angels' hands,
then fell within and round the chariot),
seen through a veil, pure white, and olive-crowned,
a lady now appeared to me. Her robe was green,
her dress the colour of a living flame.

Purgatorio, XXX

The three colours with which she is dressed represent the theological virtues; her olive crown is a symbol of wisdom and peace. With this triumphant scenography, worthy of the most renowned theatres, Dante concludes the sacred representation.

Purgatory is also the Canticle that is richest in allegories and symbols: the same mountain soaring towards the sky is a symbol of elevation through purification in the Earthly Paradise, located on its summit. There are seven frames, like the deadly sins: seven is a magical and sacred number, from the Bible onwards. Seven are the choirs that precede the Holy Ark of David's exemplum in the frame of the proud and seven are the entreaties of the Pater noster that the proud recite. Seven also were the liberal arts in Dante's time (grammar, rhetoric and dialectic in the Trivium, arithmetic, geometry, music and astronomy in the Quadrivium), and seven are the musical notes. The sacredness of the number seven is also confirmed by the Koran.

And if the number three, perfect because it symbolizes the trinity and the omnipotence of God, justifies the three Canticles of 33 Cantos each, to which the first Canto of introduction is added, the number four represents the four cardinal points, that is, the immensity of God: seven therefore unites the omnipotence and the immensity of God.

Beatrice manifests her name at verse 73, therefore, at the exact centre of the 145 verses, of Canto XXX of the Purgatory, and adding the digits of these two numbers $(7 + 3)$ and $(1 + 4 + 5)$ always gives us 10, the perfect synthesis of $3 \times 3 + 1$, that is, of the trinity and unity of God. Canto XXX is a multiple of 3 and 10, and also the sixty-fourth Canto of the poem (the sum of the digits $6 + 4 = 10$), while the number of the 63 Cantos $(6 + 3)$ preceding it and that of the remaining 36 $(3 + 6)$ both give us the number nine as the sum of their digits. And, of course, $63 = 6 \times 10 + 3$. We are in the presence of numerology, a discipline so dear to Dante, or in the presence of the cabal, as it were.

Canto XII of *Purgatory*, where the proud are found, is the one in which the number three and its multiples, already anticipated in the three examples of Canto X, in the three characters of Canto XI and in the three Cantos of the proud, acquires its maximum symbolic value. There are twelve examples arranged in three quaterns, each beginning with the same word ("vedea", "o", "mostrava") and ending with a thirteenth verse that repeats the same words. It is a real triptych, and the whole forms a sort of sestina of thirty-nine verses $(36 + 3)$; let's remember that the sestina is a poetic composition of six stanzas of six verses each, in which the same words come back in rhyme according to a strict order, with a final envoi that uses all six of the previously used words. Moreover, each example is closed in the turn of a triplet and the three quaterns seem to indicate three categories of boastful people: those punished by the deity, those punished by themselves and their own remorse, those punished by their enemies or victims.

The Golden Number

One of the fields in which Arab influence had a notable impact in the thirteenth century was mathematical research, and the most significant scientist of this turning point was Leonardo Pisano, also known as Fibonacci, who was born around 1170 and died after 1240. The number is central in Dante's cosmogony, mainly for the allegorical values assigned to particular numbers such as 3, 7 or 10. Fibonacci, rather than their symbolism, was instead attentive to their practical utility: we owe to his famous treatise *Liber Abbaci* the introduction

> of the nine digits of the decimal system, which he called "Indian," as well as the 0, in Latin called "zephirus."
>
> He had absorbed this new way of doing mathematics from the Arab algebraists, with whom he had come into contact during his stays in Algeria while following his father, Guglielmo dei Bonacci, a wealthy merchant from Pisa. He was then able to perfect his knowledge by traveling in Egypt, Syria, Greece, Constantinople and Sicily, where he came into contact with the court of Frederick II: the Swabian emperor often kept the young man to himself and used his advice for subtle questions of geometry and algebra.
>
> Fibonacci had a brilliant spirit, an absolute mastery of the discipline, an iron logic in his reasoning, a surprising originality in finding more elegant demonstrations of classical results and in proposing new problems. It was thanks to him that, in Europe, there was a happy synthesis among Greek Euclidean geometry, Diophantine's art of numbers and Arabic algebra. To estimate the growth of a population of rabbits, Fibonacci introduced a famous numerical sequence in which, starting from 0 and 1, each term F_n is obtained by summing the two numbers that precede it:
>
> $F_{n+2} = F_{n+1} + F_n;$
> $F_0 = 0, F_1 = 1.$
>
> You then get a succession of numbers that grow very quickly:
>
> 0, 1, 1, 2, 3, 5, 8, 13, 21, 34, 55, 89, ...
>
> whose ratio between two consecutive numbers tends rapidly to the golden number of the Greeks, the golden number par excellence in art and nature:
>
> $F_{n+1} / F_n \to (1 + \sqrt{5})/2 = 1.61803...$

It has also been noted that the number of verses in Cantos XIV-XX constitutes a series: 151, 145, 139, 145, 151. As we can see, there is a symmetry that has, as its pivot, the number 139 of Canto XVII, consisting of the Figs. 1, 3 and 9, which remind us of the sacredness of the number three and its multiples. Canto XVII is the central Canto of this Canticle and contains the scheme of Purgatory and the theory of love, which, for Dante, is fundamentally an expression of God, unity and trinity, while acedia, understood as a lack of love, is placed at the centre of the seven deadly sins, that is, at the centre between good and evil. In short, if the number seven is important, the number three is the very pivot of the entire Comedy, because, for every Christian, it is the main figure, foundation and purpose of the Universe and of human life.

Round Trip from Ormuz

The idea was to reach Ormuz, where the Arabian peninsula, with its promontory, creeps towards the Iranian coast, almost to the point of pinching it, and from there to embark for the Catai. It was the two Polo brothers, Niccolò and Matteo, who had imposed that decision, mindful of all the brigands, mishaps and wars they had encountered on their previous overland journey to Kublai Khan.

They had not, however, been able to avoid encountering brigands who had robbed them. The account of this unpleasant close encounter is, as usual, incisive but brief, and extremely lapidary, in spite of the loss of several lives. The brigands in question had dark skin, because they were descendants of Tartars and Indian women, and, for this reason, they were called Caraunas, which meant "mixed." These brigands conducted raids using "magical and diabolical arts to darken the day and bring darkness." Thanks to the magic of darkness, they plundered travellers of everything and then sold as slaves those who could not pay the ransom for their kidnapping: "And know that in that darkness messer Polo was also taken and ran the risk of being taken prisoner: but he managed to escape and took refuge in a castle while some of his companions were taken prisoner and sold as slaves and some were even killed."

In the account of the journey, however, there is no lack of stories with a comic vein. One of these concerns the kingdom of Cherman, made up of "good people, simple and peaceful, who help each other with great solidarity." Talking one day with the wise men of the kingdom, the king confided to them a doubt that had been haunting him for some time: he could not understand why the neighbouring kingdoms of Persia were inhabited by evil and bloodthirsty people, while he had the good fortune to govern such a mild and peaceful people. These wise men answered that the cause was in the land. The king took them literally: he sent seven ships to load into their holds the land of their neighbours and bring it to his court. When this curious cargo arrived, he ordered the earth to be spread on the floor of some of the halls of the palace and covered with carpets. These were the halls where a banquet had been prepared: upon arrival of the dishes, those present began to insult each other with outsized words and by making very eloquent gestures with their hands, the first fists flew, some drew out their daggers, and soon there were dead and wounded lying about. "So the king was sure that the cause of the difference between his people and the others truly lay in the land of their countries."

After a long journey, Marco Polo and his companions finally arrived at the coast near Ormuz, at the narrowest point of the Persian Gulf, from where the dunes of the Arabian Peninsula could be seen in the distance. A wind was blowing so hot from the land that the only remedy was to stay in the water, as the children were, in fact, doing, rolling in the sand and immersing themselves again and again, splashing water on each other's faces. Only the date palms that grew near the sand bumps provided some shade.

A few meters from the shoreline stood out the silvery color of the tall fishing nets planted in the water on long poles where a thousand seagulls circled, fighting in a perpetual rustle of wings and in the air torn by the overbearing sound of their garritos: in those places, they had been fishing quite indolently for centuries, for the waters were full of lobsters, crabs, shrimps, as well as flying fish and gobies that were easily found along the coast, so that you only had to wait, which is what the locals did. The road leading to the port opened up between rows of houses with ochre walls and tiny windows, almost slits, to shelter the residents from the intrusiveness of the sun. It felt like being in a long stretch of sand squiggles; the sun was so high in the sky that it lacked any shade.

At the port, there was great excitement along the quays: the sailors were unloading large baskets full of red coral and sea sponges, fish of all sizes and species, such as sharks, large tuna, scorpion fish and mackerel, brightly coloured octopuses with tentacles a metre long; near the ropes of the boats were piled sacks full of coconuts, pine cones and tropical fruits; in the air, there was a strong smell of incense and cinnamon. With their heads covered by rudimentary turbans and a cloth around their groins, but otherwise naked and barefoot, dozens of Arab sailors went up and down the holds without pause, shouting and waving at their nearest companions.

The Polos hoped to find a passage to India on one of those Arab boats, perhaps even directly to the distant Catai. But their hope turned to misery when they realised how those ships were made! In the bazaars scattered around the port, one could hear thousands of stories of shipwrecks that only reinforced the three Venetians' perplexity: they were used to the sturdy hulls coming out of the Venetian Arsenal, held together by two-palm long nails and tarred in a workmanlike manner; when they saw the sides of the boats there, made of rushes and held together only by coconut husks, they understood that those were hulls destined to sink at the first storm, at the first serious downpour. By coming to Ormuz, they had only wasted time; unfortunately, there was nothing left for them but to turn back, retrace the entire route by which they had come with saint-like patience and take the road to Catai that passed the highest mountains in the world.

Flowers and Trees, Light and Fire

The flowered valley of the princes, in Cantos VII and VIII of *Purgatory*, corresponds to the noble castle of Limbo, and is an oasis in Antipurgatory where those who performed important functions in earthly life find hospitality. These princes on earth did not fulfil the duties of their role, preferring the pleasures of the senses to the voice of reason: they were weak then in the face of sin and are now distressed before temptation. Dante meets here the Emperor Rudolph of Hapsburg, the King of Bohemia Ottokar, Philip III, King of France, Henry of Navarre, Peter III of Aragon, Charles I of Anjou, Henry III of England, and William VII, Marquis of Monferrato.

In this "little valley," with the words of the Salve Regina, "Te lucis ante," Dante begins—in Canto VIII—the dramatic scene of temptation that depicts the eternal struggle between God and Satan, between good and evil. The souls in Purgatory need to know the roots of evil in order to arrive at complete purification, to see temptation in order to save themselves. The soul that sings the hymn "joined his palms then lifted them, eyes fixed towards the orient," in a theatrical attitude that helps to outline a plastic picture. The hymn, attributed to Saint Ambrose, was sung by priests and the faithful during the evening hour known as Compline to ask God for protection during the night.

The scene of the temptation, which takes place only in front of the negligent princes, is also a sacred representation, staging the battle between the serpent and the angels, and has both religious and political significance. The slithering serpent ("a snake appeared, perhaps like that which fed Eve bitter food") is evidently the symbol of Satan, while the two angels represent the defence of faith and the colour of their robes—green—the hope of salvation. There are "two swords, each flashing fire, the point of each was blunt and broken off." The fire engulfing the swords indicates the terrible justice of God, while their truncation may mean that the angels act in defense, but never in offense, and that temptation can be resisted and put to flight, but never completely annihilated and killed. The fire of the swords will dispel the darkness through which the wiles of the devil spread.

The door to Purgatory, in Canto IX, is described as narrow, because it is supposed to be difficult to save oneself. The three steps of the entrance are part of the three moments of confession set by Saint Thomas in the *Summa Theologiae*. The first step, "marble, white," is the examination of conscience with which the sinners scrutinize themselves, as in a mirror; the second, "tint more dark than perse," more black than dark, is the oral exposition, with which the soul hardened by sin breaks from pain; the third, "porphyry, it seemed, and flamed as bright as fresh blood spurting from a severed vein," is the fire of charity that drives the soul to active penance through suffering and the performance of good works, which, for Dante, exemplifies entry into Purgatory. Finally, the threshold that looked like "adamantine stone" represents the firmness of the confessor and the solidity of the foundation of the Church. The seven P's on his forehead engraved "with sword point" represent the seven deadly sins, which Dante will have to get rid of. The grey of the angel's robe is a symbol of penitence, while the silver key represents the confessor's doctrine and the gold key his authority. At this point, Dante's solemn entrance is expressed by the hymn, also attributed to Saint Ambrose, "Te Deum laudamus," "as voices mixed in that sweet tune," sung and accompanied by the organ.

Another purgatorial illumination appears in Canto XV, which, depicting the third circle, the place assigned to sinners who died in anger, immersed in a dense and black smoke that symbolises the blinding of passion, is considered the Canto of Light. The Canto begins with the burst of a powerful sun ("The sun's rays struck us straight along the nose"), to which is added, "by splendour far more bright than first," "some such light that broke, reflected back from refracted light."

Compare: from water or reflective glass
a ray of light leaps back as opposite

Purgatorio, XV

This light, which dazzles even more than that of the sun, comes from the face of an angel and strikes Dante's eyes with the same force as a ray suddenly reflected by water or a mirror. The illumination produced by the sun and the angel is joined by the illuminating power of reason, which Virgil expresses in a beautiful tercet:

You thrust your mind, back down there
and, thinking still in terms of earthly things,
you tease out darkness from the light of truth.

Purgatorio, XV

In the Canto, he discusses the peculiarity of heavenly goods that can be communicated to all without dividing them up, like the light of the sun that illuminates different bodies without losing intensity. Cultural illumination joins, in Dante's journey, that of the moral and the religious, directing him towards the path that leads to God and truth. The path of light confirms and concretizes the value of the hymn to the sun that Virgil declaims in Canto

XIII, when, in the second circle, the envious, their eyelids sewn shut with an iron wire, are prevented from seeing.

We then pass from the light of the sun to the wall of fire, which Dante encounters in Canto XXVII, in the seventh frame, that of the lustful; the passage of the wall of fire is obligatory for all non-virgin souls—or, perhaps, for all souls—who go to Paradise; it is no accident that we are among the lustful, a sin with which Dante has considerable experience. Baptism by fire and burial by fire can be found in the most ancient human cultures: for example, Saint Paul and various Fathers of the Church, including Saint Ambrose, write of salvation by fire.

To sunlight and fire are added, in Purgatory, a variety of colourful vegetable presences. We have seen the flowery valley of the princes; let us now approach two very singular trees observed by Dante. The entrance and exit from the frame of the gluttons (Canto XXIV) are enclosed by two inverted trees, similar to fir trees, whose branches become shorter and shorter as they descend from the top to the bottom and feed from above. The first tree is found at the end of Canto XXII and could represent the tree of life, a symbol of immortality and communion with God, while the second tree, mentioned in Canto XXIV, appears before Dante "ripe-branched and bright": its fruit, however, is forbidden as a penance to gluttons, who stretch out their arms in vain, in a torment similar to that of the mythological king of Lydia, Tantalus. Tantalus stole ambrosia and nectar from the gods and kidnapped Ganymede, and was thus condemned to the Underworld, where he was immersed in a pond near trees laden with fruit, but was, nonetheless, fated to always remain hungry and thirsty, because the water receded when he tried to drink it and the wind blew the branches with fruit into the air when he tried to grasp them.

In Canto XXXII, Dante finds instead "a leafless tree. Its every branch was stripped of greenery": this is the tree of knowledge of good and evil, first despoiled by Adam and Eve, representing a limitation to man's knowledge and action. This tree too, like those in the frame of the gluttons, has wider branches at the top, indicating that mankind must feed on God. The tree blossoms again thanks to redemption, "in colour opening to more than violet and to less than rose," that is, with violet flowers representing the blood of Christ and the penance of Holy Week. Also in this Canto, Dante sees "the budding apple flowers which make the angels greedy for their fruit," the flowers of beatitude offered to the three apostles Peter, John and James by Christ, depicted with the apple tree of the Canticle of Canticles during his transfiguration. And Beatrice, whom Dante sees "seated on the root of that," would be the revealed truth.

In this context of illumination of the symbolism of the sacred representation, here, in the last Canto of *Purgatory*, is Beatrice's prophecy that "[t]he eagle, leaving feathers on the Cart" will not remain for too long without heirs. There will come a "FIVE HUNDRED TEN AND FIVE" sent by God, who will kill the woman with whom the giant is flirting. This is one of Dante's most cryptic prophecies. One wonders whether Dante is speaking of the same suspension of the Empire or of two different phases: the very long one initiated by the end of the Swabians and the very recent one caused by the death of Henry VII. No one has so far deciphered the riddle. Many believe it refers to Arrigo VII, since the Roman numeral DXV can be anagrammed as DVX, and, in this case, it would be Dante's homage to the Emperor Count of Luxembourg, who would still have been alive at the time of the poem's composition. Arrigo VII died at Buonconvento on August 24, 1313, but the *Inferno* and *Purgatorio* were published in their entirety before August 1313. The prophecy could, however, be meant to apply to after the death of Arrigo VII and could refer to Arrigo's son, John of Bohemia: using a cipher widely used for confidential messages from the ecclesiastical hierarchy, assuming Johannes as the key, we get that "five hundred ten and five" can be rendered with the letters f, i, e, as in "filius imperatoris enrici." Still, we don't know whether anyone will ever manage to unravel the mystery.

The Highest Place in the World

Having decided not to risk the passage to China on the perilous reed boats of the Arab sailors, the Polos had no choice but to cross the desolate highlands of Central Asia, in particular, the Pamir, the "roof of the world":

> *Continuing in the direction of Eastern Greece, after having walked twelve days, one has to count another forty always in the same direction through mountains and valleys, passing rivers, crossing deserts and never meeting houses or vegetation [...]. This district is called Belor and its inhabitants live high up in the mountains: they are idolaters and savages, they live only by hunting, they dress in animal skins and are very wicked people.*

Crossing the Pamir massif (and the adjacent Belor district) was no joke: the extreme climatic conditions, with temperature ranges of several tens of degrees from day to night, had always made the place particularly inhospitable. Even the fire had a different colour and did not cook as well. The region was divided into two large areas: the western one, dominated by imposing mountains and deep valleys crossed by streams rushing down from the surrounding glaciers, and the eastern one, made up mainly of high altitude desert plains and bordered by salt deserts. At lower altitudes, the landscape was dominated by maples, a few wild apple trees, and junipers and birches; along the streams, one could see willows, small forests of poplars and hawthorns, and a few oaks. However, the higher you went, the more the silhouettes of the trees disappeared from view, giving way to an enormous tundra that stretched to the end of the world.

Those desolate lands were the kingdom of the argali, the wild sheep with the longest horns, heavier and more twisted than those of any other sheep species: "With these horns—Marco Polo noted—the shepherds make large bowls in which they eat; and they also make poles for the fences with which they fence off the places where they keep their livestock at night." Additionally, they were places infested by wolves and snow leopards, which, besides the sheep, fed on Siberian ibexes, deer, hares and goitered gazelles. Every now and then, the snout of a marmot would appear in the bushes, and you could occasionally make out the cunning eye of a lynx.

Over the years, long caravans of sturdy Bactrian camels laden with silk, pilgrims, wayfarers and Buddhist monks had passed through the gorges of those mountains, leaving testimony in their writings of the glacial cold and strong winds of those places. Some of them swore that they had seen traces of dragons and crocodiles. Marco noticed a large plateau between two mountains with a lake out of which flowed a beautiful river with water as clear and limpid as crystal, surrounded by meadows so rich that a thin beast would grow fat after only ten days in that pasture.

The Polos spent a few days with the Kyrgyz shepherds, sleeping in their yurts and enjoying yoghurt made of goat's milk, eating chapati made of rye cooked between two layers of dried camel dung, the only fuel in that place without trees or bushes. They were a people of Turkoman origin who, like other Turkic peoples, had been subjugated by Genghis Khan. Their close familiarity with the glorious people of the steppe had meant that the Kyrgyz had also acquired distinctly Mongolian somatic traits, becoming indistinguishable from their invaders. They also lived by trade and the arts, moving among the few castles that stood in those valleys. Marco also noticed the presence of churches and Nestorian Christians who spoke an idiom different from that of the rest of the population. In those remote places, at nightfall, when the fires were lit for warmth, humanity and nature became one, and the silence of the night was broken only by the whistling of the wind and the howling of wolves.

Coming down from the Pamir mountains proved to be even more dangerous than going up: along those icy paths full of sharp stones, the animals lost their balance and the camels swayed dangerously on the edge of those ravines whose bottom was not visible. Sometimes, a horse would stumble and throw its precious load onto the ground, after which there would be curses, and then, with steely perseverance, everything would be loaded back on and they would continue their journey. Wading through icy torrents, holding the animals by the muzzle along those endless sassafias, scrutinizing the patterns of the constellations in the night; after forty days of this ordeal, Marco and his travelling companions finally reached Great Turkey, a region governed by the powerful Caidu, the fiercest adversary of Kublai Khan.

They passed through the city of Cascar, an obligatory stop on the way to the Catai and the crossroads of many camel routes coming from Kashmir. To the south, many days' walk away, was the noble city of Samarkand, a millenary junction of the Silk Road, the domes of its splendid madrasas covered with tiles the colour of the sky. Alexander the Great also visited it in 329 B.C., and was fascinated: "Whatever I have heard about Samarkand is true, except that it is even more beautiful than I had imagined." Marco Polo didn't have the chance to visit it, but he nonetheless didn't hesitate to tell a tasty legend about a miracle that took place right there, in the church of Saint John the Baptist, succeeding, as usual, in transforming true historical facts into a grandiose fresco of fairy-tale medieval Asia.

At the time of the conquest of the city by Ciagatai, a Mongol leader and governor who was the brother of the Great Khan Ogodei, the basilica dedicated to Saint John was built to celebrate his adherence to the Christian faith, using a large and beautiful stone that had belonged to the Saracens. This stone became the base of a column that was placed in the middle of the church to support the roof. Things became more complicated when Ciagatai died and was succeeded by his son, who had recently converted to Islam. The Muslims then demanded the return of their stone, refusing any price offered in exchange by the Christians. Dismayed, the Christians appealed to the glorious Saint John, who, in fact, answered their prayers: on the morning of the day fixed for the restitution of the stone, the column resting on that stone rose into the air, leaving a space of three spans, and there it remained, before the eyes of all, even when the stone was removed by the Saracens. The miracle, said Marco Polo, aroused great wonder, and the news of it spread like lightning throughout the East.

The Elements, the Ether and Transhumanar

Dante's journey so far has taken the form of a strenuous and exhausting walk through Hell, always on the left, and an ascent to the right that has become increasingly gentle in Purgatory. Now, Dante "changes pace," radically transforms himself and begins an extraordinarily visionary flight, for the knowledge and technology of his time. At noon on Wednesday, March 30, 1300, he begins his journey to Paradise, thanks to Beatrice, who guides him. He is in the element of fire and is about to leave the sublunar world.

The sublunar world is composed of four elements: earth, water, air, and fire. In the thirteenth and fourteenth centuries, following the authority of Aristotle, it was still accepted that these elements tended to stratify naturally according to their different degrees of density: water above earth, air above water and fire above air. We learn from the *Convivio*, and we find again in Canto XVIII of the *Purgatorio*—"as fire moves upwards to the heights (by virtue of its form, it's born to rise to where it may, as matter, most endure)"—that the determining force of this separation and of the relative movements is love, understood as an attractive virtue, comparable to that which is manifested in magnets.

For Dante, and for Aristotle, these elements result from the two-by-two combinations of the four properties—cold, heat, humidity and dryness—from which all bodies, human temperaments, ages, seasons and climates derive. In the celestial spheres, there is no fire: it predominates in the upper part of the atmosphere, as does air in the lower part. The ethereal matter of the heavenly spheres is purer than the four elements, and circular motion is natural to it. In the upper part of the atmosphere are produced the shooting stars and meteors, originated by hot and dry evaporations, which, rising, are dragged into the rotation of the atmosphere and burn.

To describe this new divine world, Dante appealed to Apollo, the god of Wisdom and personification of divine inspiration, in order to address the lofty matter of Paradise, and thus merit the poetic laurel. The proem to the third Canticle extends for thirty-six verses, and is thus three times as long as the proem of Purgatory and four times as long as that of Hell. The greater breadth and solemnity corresponds to the importance of the subject matter. Let us remember that Dante now describes the "holy realm" as never before. Dante's poetry must be inspired by God and not attempt to compete with divinity in the representation of what exceeds human limits.

We are in the spring equinox and the sun is in conjunction with the constellation of Aries, which makes its rays more beneficial to the world. Dante imitates Beatrice, turning to stare at the sun like an eagle: "and fix my eyes—beyond our norm—straight at the sun," just as a reflected ray rises at the same angle as the first ray, without affecting his vision, so the Poet can stare at the sun more than he would on earth.

At this point, the miracle of Dante's "transhumanisation" takes place, a process that goes beyond the human condition, a phenomenon that cannot be explained rationally: "To give (even in Latin phrase) a meaning to 'transhuman' can't be done." Dante has here opened the door to the beyond-man and the superman, well before Friedrich Nietzsche and Gabriele D'Annunzio. The Poet loses his appearance and undergoes a transformation:

what Glaucus, tasting grass, was made to be,
consorting with the other ocean gods.

Paradiso, I

Glaucus, mythical fisherman of Boeotia, was the son of Poseidon, god of the sea, and of a nymph of the Naiads. According to the legend, presented by Ovid in the *Metamorphoses*, Glaucus one day placed his fishing net on a meadow and saw that the fish, eating the grass that covered it, came back to life and threw themselves back into the sea. Glaucus tasted that magic herb and became immortal, his legs turning into the tail of a fish. We note that Dante will often resort to mythological similes in the Cantica to represent situations lacking earthly terms of comparison, as in this case.

The Poet cannot tell whether he is still in possession of his mortal body or if he had been reduced to his soul alone, but he fixes his gaze on the heavens, which revolve to the tune of a harmonious melody, and it seems to him that the light of the sun illuminates all of the surrounding space in an extraordinary way. He asks Beatrice about the origin of the sound and light. She explains to him that she is no longer on earth and is ascending into Paradise, quickly overcoming the sphere of fire that is lower down. How is it possible, Dante wonders again, that, endowed with a mortal body, he is ascending beyond air and fire? Beatrice heaves a deep sigh and, as a mother would do with a son who speaks nonsense, replies:

There is an ordered ratio.
between all things there are. It's this – such form –
that makes the universe resemble God.

Paradiso, I

All creatures are inclined towards God according to their nature, and tend to different ends by different paths, according to the impulse that is given to them. God resides in the Empyrean, and Dante and Beatrice head there because their natural instinct drives them towards their principle, which is God. Beatrice's theological explanation anticipates the stylistic features of much of the Third Canto, in which Dante's scientific doubts are resolved with doctrinal arguments and it is reaffirmed that human philosophy is insufficient in itself to understand the mysteries of the Universe.

Dante must believe that he is ascending with his whole body into Paradise, not being able to comprehend it. Dante's journey towards the light is therefore his journey towards God, marked from the very first verses of Canto I by the divine glory that reverberates throughout the Universe, starting from Empyrean, "which takes from Him most light."

The Silk Road

According to an ancient Chinese legend, the birth of silkworm cultivation is due to a fortuitous discovery made in the mists of time by Empress Xi Ling Shi, the very young wife of Emperor Huang Di. In fact, around 3000 BC, the empress was sipping hot tea in the shade of a mulberry tree when a silkworm cocoon fell into her cup. Annoyed, she angrily grabbed the cocoon to remove it from the tea, but, due to the heat of the drink, the little white wad magically began to slip through her fingers!

The very long thread swung in the air without breaking and, swaying, shone cheerfully in its brightness. It immediately occurred to her that this mysterious thread could be woven, worked, collected, dyed in a thousand colours… When she finally managed to remove the cocoon, she noticed that there was a worm hiding inside, the same worm that loved to feed greedily on the mulberry leaves in her garden. Intrigued, in the following days, she decided to study the thing more closelt, thus coming to understand how the caterpillar, once it had reached maturity, was forced to wrap itself completely in the cocoon of that white slime—which he produced endlessly from two small holes near his mouth—to give rise to his metamorphosis, first into a chrysalis and then into a splendid butterfly.

Ling Shi waited for the first opportunity to ask the emperor for permission to plant more mulberry trees so that she could cultivate these precious worms. She also had looms built to spin that wonderful natural thread and taught many women the noble art of weaving. Ling Shi also became the first woman weaver in history, at least, according to legend.

In ancient China, in addition to the emperor and empress, always wrapped in elegant yellow silk fabrics, people of the court and priests also began to dress in silk; it had soon become a luxury good. The emperors were careful to jealously guard the secrets of its cultivation and processing: they issued edicts so that anyone who revealed how this amazing thread was obtained was put to death.

From the villages of southern China, those graceful fabrics began to travel on the shoulders of mules or on the humps of camels along the commercial arteries that united the Far East with the West. The Parthians, a population settled on the Persian plateau, fell in love with the soft and refined consistency of that mysterious fabric, marking the first time in history that an ostrich egg was bartered for a roll of silk. The Roman world got to know silk thanks to the colourful banners carried by the Parthians during the disastrous defeat that they suffered in the battle of Carre in 53 B.C.; from then on, Roman senators, generals and consuls developed an expensive mania for that delicate cloth, striving to obtain it at all costs in the markets of the cities that opened onto the Mediterranean, where silk soon became almost more precious than gold.

At the fall of the Western Roman Empire, Rome was supplanted by Constantinople, and one of the Byzantine merchants' pressing objectives was to establish contacts with China, where the light-as-air fabric came from and whose origin, however, remained mysterious. But, as we all know, there is no secret that is destined to last forever: another legend, in fact, tells that, in 552 A.D., two Nestorian monks from China who had silkworm eggs hidden in the hollow of some reeds revealed the secret of silkworm cultivation to the Emperor of Constantinople, thus giving rise to what became one of the most flourishing sectors of the Byzantine economy.

Silk was obviously not the only good that arrived in the West from the heart of China via the caravan routes. Together with the fabrics of sparkling colours, into the markets of Antioch, Constantinople, Alexandria, Palmyra, Petra, Smyrna and Damascus came the lapis lazuli of the Karakorum, the perfumed moss gathered near the high mountain lakes, the spices of the Indies, the honey of the chestnut woods behind the Oxus River.

The Silk Road is one of those evocative names in which images of medieval travellers, missionaries with sun-baked faces, slow caravans of camels along paths dominated by snow-capped peaks, with sacks full of bejewelled jugs, coloured carpets, precious stones, pink salt, pomegranates and dried dates loaded on their humps are all mixed together as in a luminous and gigantic kaleidoscope. That name, so romantic, was coined for the first time in 1877 by the German explorer and geographer Ferdinand von Richthofen, even though the Chinese had had their own name for that web of roads, mule tracks and paths that connected the villages near the oases, the villages where the various shepherds went to sell or buy their sheep, to exchange bunches of grapes for the sweet melons of the valleys, and to push on towards the West, since at least the first century A.D.

Having crossed the mountain passes of the Pamir, the roads were dispersed in a thousand rivulets: there was the one leading to India, the one continuing towards the Iranian plateau, and then there was a whole network of dusty paths leading to the basins of the Tigris and Euphrates Rivers. The most ancient section was the eastern one, which, starting from North China, near the oasis of Dunhuang, forked into two main routes—one following the basin of the Tarim River, the other the edges of the fearful Taklamakan desert—before finally reuniting near Kashgar.

Along those roads, time was indefinite, unchanging and silent. Those paths had seen the passage of soldiers in search of gold and glory, merchants thirsty for riches, thieves and swindlers, cunning adventurers, sellers of horses and stuffed birds, astute diplomats and disillusioned fortune seekers. As on every transit route, along the caravanserais scattered in the valleys and in the villages at the edge of the deserts, there was a mingling of people and their stories: one could meet blue-eyed Chinese, perhaps descendants of Alexander the Great's soldiers or of the centurions of the Roman general Crassus, and women with moist purple eyes; one could hear intertwined dialects of Persian and the ancient language of the Mongols, a Babel of languages and sounds.

At the time of Marco Polo, the Silk Road was busier and more active than ever, and remained so until the fifteenth century, the time of the pax mongolica imposed by Genghis Khan. The Mongols used those routes to buy large quantities of tea, a drink that they had come to know and drink with pleasure after conquering China. The path of those routes travelled by camel drivers passed through the most inaccessible regions of the Earth: the arid deserts were followed by incredibly dangerous high peaks; the steppes stretching as far as the eye could see were contrasted with the impetuous waters of the rivers that flowed directly from the glaciers.

Marco Polo's uncle and father had probably travelled along those routes on their first journey to the East and were appealing to their memory to find the road that would lead them back to Kublai Khan. They had decided to follow the Silk Road along its southern artery, which started from the lands of Badakhshan, a region rich in precious rubies, the highest quality blue lapis lazuli and sapphires the colour of the sea. But, immediately after leaving, they had to deal with the fact that Marco had fallen seriously ill, forcing the whole group to take a break that lasted more than a year. Perhaps it was malaria: it was the shamans of the village where they had stopped who, at a certain point, advised his father to take young Marco high up into the mountains, where he could enjoy fresh, pure mountain air that would "sweep away evil spirits." They did so, and Marco effectively recovered after a short time, allowing them to resume the journey to Catai.

Those desolate lands were often infested with straggling soldiers and groups of thieves, and to protect themselves, the locals preferred to take refuge in fortified dwellings at the mouths of the valleys. The rooms of those small manors were filled with soft carpets in the colors of fire and the sky. The nearby bazaars sold wheat, pistachios, almonds and, especially, rock salt, used to preserve meat. Children enjoyed hunting porcupines and hares that popped out of holes in the stone-filled ground.

Most of the people Marco met at the beginning of the Silk Road turned out to be Muslims: Marco called them "Saracens" and often described them in unflattering terms as cutthroats, thieves, murderers and even drunkards, which must have been a serious offence to those who were supposed to follow the Prophet's teachings (although the smell of the excellent vin cotto, wine reduced to a kind of syrup, often got the better of any good intention). Going towards the East, Marco also came across more and more "idolaters," i.e., Buddhist or Hindu devotees. Among them were monks who wore tonsures and were skilled in the enigmatic art of magic: they were able to perform incredible wonders, walk on fire, enchant cobras, darken the day or levitate in the air. Marco was enchanted by the beauty of the Kashmiri women, by their silky long hair as black as night, by their big purple eyes, by their soft dark skin.

Along the way, the Polos also met some Nestorian monks who gave them news of the great leader of the Mongols and several patrols of Kublai Khan, out on patrol. They often stopped to sleep in the mail stations, which were clean and offered food and water. There, they learned that Kublai Khan had long known of their arrival and was looking forward to seeing them. Here and there, scattered along the valleys, were a few castles with crenelated towers the light satin colour of the sand: at nightfall, the fires lit on the walls of the forts increased the magic of those places, while, in the distance, you could see the red foxes running away and hear the howling of wolves.

From time immemorial, those inhospitable places were one of the most important centers of Buddhism, as evidenced by the numerous chapels carved into the rock and their walls adorned with paintings, as well as the presence of monasteries full of idols where rich sacrifices were made to the gods. There were also shamanic rites concerning the dead, whose bodies, soaked in camphor and covered with resin, were burnt: the precise date of the burning ceremony was decided by astrologers, who calculated the most favourable position of the constellations, according to the date of birth of the deceased.

They also calculated in which direction the corpse had to leave the house to prevent the deceased from turning into an evil lemur, capable of harming the surviving inhabitants of the family: it often happened that this direction did not coincide with the front door, but, in this case, they would not hesitate to knock down the walls to create the right opening.

Before arriving in Catai, the Polos had to face one of the most difficult trials of their journey: crossing the fearful Gobi desert. At the extreme edge of this great ocean of sand was the town of Lop, where the Polos rested for a good week, stocking up on water and provisions: they loaded the mammoth Bactrian camels with dried dates, pomegranates, almonds, fruit, and smoked meat, and ventured out on the crossing that was to last about a month.

They had heard the strangest and most magical legends about those places devoid of all living things. The most disquieting story concerned the "talking sands," that is, the voices of spirits that bewitched those who, travelling at night, lost sight of their companions and became so disoriented that they could no longer find their way and disappeared among the dunes: "There have been men," wrote Marco, "who, passing through this desert, they saw an army of people coming towards them with impetuosity, and fearing to be attacked and robbed, they fled, abandoning the main road; but then they never found it again, and miserably starved to death."

However, there were not only the voices of the night: during the day, you could also hear the distant rolls of drums, sounds of trumpets and shouts of people. To avoid getting lost in those expanses of sand, travellers would always go in companies, even putting bells around the necks of the animals that, through their jingling, helped them not to get lost. When the parties woke up in the morning, the sand would crunch under their feet, due to the condensation that had formed during the night. The guides who accompanied the Polos scanned the horizon in search of oases where they could take shelter and possibly find a water hole for the animals to drink from. Everything was extreme around there: the reflection of the sun, the temperature, the endless expanse of sand, the desire to drink. Always following the direction of the stars of the Big Dipper in the sky, the Polos crossed that sea of sand and finally reached the fabulous Catai, the land of the Kublai Khan.

Light, Dance and Music

Light and music pervade Dante's extraordinary journey through Paradise. In many places in his works, Dante confided thoughts, analyses, considerations, and critical views on the problems of the science of light, the eye, and vision. He understood light as a movement through the air, showing that he accepted the conception of Aristotle, who considered it an immaterial energy that propagates in the same way as moving air and water, as we now know happens with sound.

According to Dante, the eye receives the aerial forms of things like a mirror and the visual faculty transmits them from the pupil to the brain, which perceives and becomes aware of the object seen. From the pupil, the visual spirit instantaneously transmits the image formed in the eye to the front part of the brain, where the sensitive virtue is located, that is, the act of conscious vision. The visual spirit, according to the physiology of the time, was the transmitting energy of visual power from the eye to the nervous center. For Aristotle, the visual process is twofold: at first, it occurs in the eye, and later in the intellect. But Dante more perceptively distinguishes it in three stages: sensation in the eye, very rapid, instantaneous conduction through the optic nerve to the brain, and cerebral representation in the imaginative faculty.

Sunlight, though of one quality and coming from one source, is received differently and modified by various bodies. Some, such as gold and certain stones, under the sun's rays, become very splendid (we would say today on account of the reflection and refraction of the light that strikes them). Other bodies—and here Dante follows Albertus Magnus—allow themselves to be crossed by light through their transparency and diffuse their colour to this and to the other objects illuminated by it. Still other bodies, such as mirrors, become so reflective in the light that they dazzle. Finally, others, such as the earth, reflect little light and diffuse and absorb much of it. In short, Dante, and Aristotle with him, understood light as a form of energy, an immaterial form in motion from the light source to the eye, or from the light source to the various bodies and from these to the eye.

He explains this well in two places in *Paradise*. In Canto XXVI, in which he finds Adam telling him about the time of his happiness and unhappiness, Dante describes himself in the moment of a sudden awakening due to a sharp and intense light that dazzles him, to the point that he cannot clearly distinguish what he sees, undergoing the sensation without being able to transform it into a distinct perception of the sensible object. And in Canto XXX, when Dante describes his ascent to the Empyrean, in a river of light that makes the Candida Rosa of the blessed appear to him, the Poet presents this dazzling apparition of "living light":

As lights, when flashing suddenly, disperse
the spirits of the retina, and rob
the eye of seeing even strong, bright things,
so, bright around me, shone a living light

that left me, baby-like, in swaddling weaves
of brilliance, so that nothing showed to me.

Paradiso, XXX

In *Paradise*, Dante is also attentive to the reflection and speed of light. On the reflection of light, we find two examples:

As any falcon's searching flight will dive,
then strike back up, or else like reflex rays,
which, angled from the first, return on high

Paradiso, I

The second ray returning from that depth
reflects like coloured images in glass –
when glass has lead plate hidden at its back.

Paradiso, II.

Finally, he also expresses his opinion on the speed of light:'

Light rays that enter amber, crystal, glass,
display such luminescence that, from when
they reach, then are there wholly, there's no pause.

Paradiso, XXIX

His optical expertise and his thoughts on the phenomenon of brightness here are simply remarkable, although he does not distinguish reflected light—produced when a beam of light hits a mirror and "bounces" off of it, i.e. the direction is opposite to that of the direction of origin—from refracted light, a phenomenon that occurs when light passes through the separating surface. Between two transparent substances, such as air and glass, the incident beam is deflected from its initial direction. The light in Paradise is intellectual light, the visible appearance of divine perfection, infinite and simple, having its source in the Empyrean and, reflected in the higher intelligences, giving light to the Universe, sparkling in the Sun and the stars.

Moon Spots

Even observed with the naked eye, the Moon appears to be a celestial body with strong luminous variations. Within Aristotelian cosmology, which expected celestial bodies to be part of the realm of perfection, and therefore completely free of irregularities, these anomalous aspects of the Moon were a real problem. The dark regions of the surface, the so-called moonspots, were explained in the Middle Ages by the lunar density varying from point to point, so that, even though the Moon had the expected spherical shape for celestial bodies, it actually appeared quite jagged.

This is the thesis that Dante proposes in Canto II of the *Paradise* when he is in flight around the Moon with Beatrice. At that time, there were, of course, no telescopes, the advent of which brought down once and for all the concept of the perfection of the celestial bodies: it was Galileo, in fact, who, centuries later, came to understand, thanks to his acute observations, that the dark regions are nothing but shadows cast by the mountain ridges and craters on our satellite.

Dante's words are answered by Beatrice, who refutes his thesis and proposes her own cosmological and cosmogonic explanation in which the initial cause is obviously God, creator of form and matter. According to Beatrice, the reason for sunspots cannot be the greater or lesser density of regions on the satellite, because, if that were the case, in solar eclipses, the rays would pass through the thinner layers, which they do not: a reasonable objection. Her proposed thesis, which, in the end, reduces to stating that the varied brightness of the celestial bodies has as its cause the angelic intelligences that distribute, from the First Mobile placed in the Empyrean, the universal virtue to the multitude of the lights that dot it is definitely more esoteric.

Leaving aside any claim to scientific value, it is important to emphasize the importance of the moonspot discussion within the *Comedy*: it denotes, in fact, the attention paid by Dante to aspects of nature and his consistent cosmological vision in which the Prime Being, God, is the cause of the marvelous diversity of the world.

Light is often mixed, in Paradise, with dance, music and song. All those who have detailed Dante's biography, starting with Boccaccio, have highlighted his great love for music. In the Middle Ages, music was considered a consequence of the great harmony that governed all of creation, and therefore, together with numbers, was explained as the natural expression of the divine. Music therefore accompanies the spiritual journey from sin to purification and redemption, and it is through music that Dante addresses the theme of God's love.

In *Paradise*, there is a scenario of accomplished musical polyphony. Dante here highlights the central role of music, along with that of light and movement. Movement derives from desire and love for God, while music and light are the propagation of God. Paradise is a world made up, above all, of light and harmonious sound, where sound and movement seem to merge and dance becomes the source of light and sound. Dante refers to the harmonic combination of several voices when he describes the following, in Canto X:

so in its glory I beheld that wheel
go moving round and answer, voice to voice,
tuned to a sweetness that cannot be known,
except up there where joy in-evers all.

Paradiso, X

In its orchestrated choreography, the dance in the *Comedy* can be understood as a movement of redemption. The different movements describe the emotions of the Poet's fellow travellers: the rhythmic chorale becomes a poetic and performative representation of the path of salvation.

The terms "ridda," "tresca" and "carola," well known to dance historians, linguistically accompany and represent the different states of mind of the protagonists of the *Comedy*. At the same time, they define the places in the poem, literally "written" by the movements of the souls that the Poet encounters while expiating his own sins or eternally rejoicing in divine proximity. And the powerful images of the dance of the stars allow Dante to redeem the classical sense of movement as a reflection of perfection.

The narrator also makes another pre-eminent performative place his own: that occupied by the dancing virtues. He thus redeems the dance from the anathemas of moralists and theologians, snatching it from the clutches of the devil and returning it to its ritual context. Prudence, Justice, Fortitude and Temperance dance in a circle and form the bridge that allows the Poet to pass from the still too human world of punishment to the ethereal and spiritual world of Paradise.

Canto XIII of *Paradise* focuses on the dance of the wise men, a dance that finds its raison d'être in the fact that Wisdom is also a dancing woman, but, above all, because, arranged in a circle, the chorus of wise men becomes an allusion to the theological concept of perichóresis (circumincessio). It is a figure that explains the nature of the trinity as eternal movement and the emanation of sapiential light and that takes on the features of an everlasting dance.

The sublimation of courtly love into mystical love marks the metamorphosis of the dance into a spiritual carol, and also allows us to overcome the apparent contradiction arising from the comparison between the chorus of the wise spirits and the vision of a group of "non-dancing loose" women. In order to adequately describe the rotating movement of the two crowns of wise spirits, Dante invites the reader to imagine the fifteen brightest stars in the vault of heaven, then the seven stars of the Big Dipper, and, finally, the two lowest stars of the Little Dipper: if these twenty-four stars formed two concentric crowns, rotating in opposite directions, they would give a faded image of the two crowns of the blessed dancing and singing around the Poet and Beatrice. The song sung by the blessed ones praises the trinity and the dual nature of Christ, human and divine.

The Canto constitutes a didactic parenthesis, since it is largely dedicated to the doctrinal question of the true nature of Solomon's wisdom. Dante tackles the very delicate question of the limits of human wisdom with respect to divine judgement, which, on the one hand, is linked to his often evoked intellectual "misguidance," and, on the other, anticipates the great theme of divine justice that will be dealt with in the Heaven of Jupiter.

The power of the *Commedia*'s gestural laboratory is rooted in the complex debate on the role of dance in the Western Middle Ages, but it is also projected into the art and performance of the subsequent centuries. *Paradise* reveals a world in which the institutional links between sounds and words no longer have any reason to exist and creates a new scenario, in which song does not refer to the usual forms of liturgy, but to light and movement. Dante says to Beatrice, interweaving vision and voice:

Why, then, does not your voice (which so delights
these spheres with that same song from holy fires
who make themselves a hood of six great wings)
bring my desires the satisfaction due?
If I in-you-ed myself as you in-me,
I would not still await what you might ask.

Paradiso, IX.

In Canto X, in which Saint Thomas speaks in the sky around the Sun, the choice of polyphony is evident, with the crown of the blessed: the voices chase each other, alternate, reunite; meaning is secondary compared to the phantasmagorical combination of sound, light and movement. In this new dimension, the human word is heard less and less. Dante does not have to ask: Beatrice and the other blessed ones read his mind and anticipate his desires. The usual distinctions between the senses also disappear: sight and hearing merge more and more frequently, and little room is left for singing and strictly auditory enjoyment.

And, in Canto XIV, in the sky around Mars, the trinity is presented in a sung and intertwined melodic movement, in which the words, almost like detached notes, echo each other in reverse order. Sound and movement are confused, and the dance becomes a source of sound and light. Thus, the undisputed superiority of vocal music accompanied by instrumental music, compared to individual vocal or instrumental expressions, is fully expressed.

In the last Canto of *Paradise*, Canto XXXIII, when the end of the journey approaches, even the music becomes quiet. In the last moments, those that precede the vision of God, the sounds return to a place of absolute remoteness: it is the sight of the mind that performs the last act, in a scenario of sidereal space, "by sudden lightning bringing what it wished." And, at the end, there remains the cosmic silence, similar to that which preceded the birth of the Universe. That silence that Leopardi will echo in another form at the end of the *Canticle of the Sylvester Rooster*: "Of the whole world, and of the infinite events and calamities of created things, not even a vestige will remain; but a naked silence, and a very high tranquillity, will fill the immense space."

At the Court of the Great Khan

After crossing the Gobi desert, Marco Polo and his companions continued eastward, soon after arriving at the province of Chienchintalas, famous for its salamander, the name by which asbestos was then known, the mineral with which the surrounding mountains abounded and which was spun to make clothes: when they were dirty, these clothes were laid within a burning fire and would immediately turn white again. The next province was Succiu, a land rich in rhubarb, where men worshipped golden idols and could have up to thirty wives.

And, finally, after more than three and a half years of travel, the borders of the coveted and noble Catai! They arrived in sight of the city of Shangdu, the summer residence of Kublai Khan, a place where the emperor loved to spend the hottest months of the year resting and hunting. A platoon of armed men on horseback came out of the city gate to meet them and escort them to the Great Khan. They were Mongol warriors, with leather armour, bows slung over their shoulders and quivers full of arrows. Niccolò Polo showed the patrol leader the golden tablets they had first received from Kublai Khan as a pass through all of the territories of his kingdom and the soldier nodded complacently. From a small hill overlooking the valley, they saw the city open up before their eyes in all its magnificence: along the walls, for about fifteen miles, there were watchtowers and troops of soldiers with spears and bows, while, inside, they could see warehouses and shops, surrounded by groves and large meadows where deer, roe deer and fallow deer roamed.

There was also a large encampment of circular tents lined with multicolored carpets that, smeared with pale earth and bone dust, shone in the sun like so many shells. Everywhere you turned your eyes, you saw banners the color of gold, flags and insignia, long yak horns raised on sturdy slats adorned with ponytails. You could see soldiers practicing with their bows or fencing, there were those who tended and groomed the horses, while the women were busy working with butter and lighting bonfires. In the midst of the flood of tents stood an imposing marble palace, a building used for ceremonies and receptions; the emperor, however, a lover of the tradition of his ancient nomadic people, lived in a palace made of bamboo canes held up by hundreds of silken ropes pulled taut. The chief of the patrol who escorted them told the Venetians that, in that place kissed by the gods, the Great Khan loved to hunt with a falcon, of which he owned more than two hundred specimens. At times, he also hunted with a tame leopard that he carried on the back of his horse, throwing it suddenly to catch his prey. Entering the city, the Polos rode down Shangdu's main avenue to their assigned tent and, as they passed, they noticed people bowing their heads in respect. They barely had time to clean and change before they were summoned to court. Niccolò instructed Marco not to say a word unless questioned, and to let him conduct the conversation with the emperor. They were received with all honours by a number of officials, bowing, making slow gestures with their hands, bowing again: shortly afterwards appeared a ceremonial man of royal bearing, who begged them to follow him to a great door, and then made a sign to enter. They stepped over the threshold, and, with deep deference, advanced with bowed heads to the middle of the hall, where they knelt and prostrated themselves with their foreheads to the ground.

After a moment's silence, the Khan's strong voice arose: "You may rise. Who is the young man who is with you?" Niccolo replied, "He is a man who will be faithful to you, and he is my son. His name is Marco." Standing up to approach the emperor, Marco at last saw the Great Khan, about whom his father Niccolò and his uncle Matteo had always spoken with great enthusiasm. He noticed that the Khan had beautiful eyes and black braids that came down in front of his ears: he was dressed in sumptuous clothes and sat on a gigantic throne of inlaid wood, all surrounded by drapes of yellow silk. Marco approached that pedestal to give him the oil from the Holy Sepulchre of Jerusalem, so jealously guarded during all the vicissitudes of that long journey. When the old mother of the Great Khan, a Nestorian Christian, heard of this gift, she was very happy.

This was the beginning of a partnership that was destined to last seventeen years, between that sixty-year-old great emperor of Asia, the prince of princes, the most powerful man in the world, and that twenty-year-old man who came from afar and showed an unparalleled intelligence and quick wit. A unique bond, almost filial, shaped by Marco's colourful storyteller's language and the

Khan's innate curiosity about the wonders of his Empire, those corners of the world that he had conquered with arms but had never been able to visit, because he was held up at court or engaged in other, newer wars.

Welcomed like a son, Marco proved to be a faithful and precious advisor to that old monarch: in a short time, he was able to learn Mongolian customs and traditions, imperial rites and ceremonies, and became fluent in the thousands of languages spoken in that oceanic kingdom. He was thirsty to know, to understand that world so far away from his beloved Venice, and so he watched and wrote down everything in his mind, silently, discreetly. It was a universe teeming with the most diverse people, the most unlikely things, the most amazing numbers: he learned there, in the ancient Catai, the astronomical figures for which he later became famous.

He noticed, for example, how, gravitating around the emperor, there was a colourful and cosmopolitan court made up of thousands of vassals, among which were dozens of Tibetan magicians and sorcerers, top experts in spells and divinatory arts who were entrusted with the task of predicting the future, driving out demons from the path of life of the Great Khan and preventing the rain from ever falling on the roof of his palace. These magicians spent their time preparing their concoctions and scanning the sky; they were also renowned for the prodigious scenes that they produced at banquets, when, on their orders, cups of wine would levitate off of the tables and move from one end of the room to the other, amidst shouts of surprise from the diners.

He discovered that Kublai Khan had four wives and many children. Each wife enjoyed the title of Empress, had her own personal pavilion, and a huge court of damsels and eunuchs. But these four wives were not enough to appease the emperor's voracious sexual appetite. For this, he turned to the Mongolian tribe of the Ongirati, known for the beauty of their girls, from whom, in the past, Borte, the favourite bride of Genghis Khan, had been chosen. In honour of that ancient choice made by the founder of all the Mongol kings, and as a good omen for the future, Kublai had his officials carefully select his bedfellows only from among the young women of this tribe: those who passed the very strict controls, during which their virginity, the proportions of their face and body, their breath and the silence of their sleep were tested, were finally admitted to the harem of the Great Khan and accompanied him in the evening to the pleasure pavilion. The others remained at court as ladies-in-waiting or weavers, sometimes even managing to marry some rich official or leader.

No one was surprised by the emperor's sexual habits: indeed, every city under the rule of the Great Khan had the famous houses with red lanterns above the entrance doors, in whose rooms the oldest profession in the world was practiced… but only until the nightly tolling of the bells, after which anyone caught in the street by patrols of soldiers was taken to prison and beaten. The best ladies of pleasure were said to be in Hangzhou, a graceful city on the water overlooking one of the most spectacular complexes of lakes and canals and graced by dozens of bridges: the renowned women of Hangzhou were not only experts in dispensing caresses and honey, but were also renowned for having refined the art of speech, entertaining and comforting their guests with sweet and reassuring words.

Marco also learned very early on that the society shaped by Kublai Khan was bloody, colonialist and rigidly compartmentalized, a legacy of the fierce years spent in battles and wars of conquest. There were essentially four groups of subjects in that kingdom.

The Mongols belonged exclusively to the first group, but they were divided internally according to their noble lineage into White Tartars and Black Tartars: only those belonging to this first group could aspire to become generals, to be appointed governors of provinces or to hold all other important elective offices. The law was extremely clear about their privileges: for example, if a Mongol were to kill a Chinese, the most he would face would be a fine, but if a Mongol were to be killed by a Chinese, the Chinese would face the maximum penalty, death.

To the second social group belonged, more or less, all those ethnic groups considered "friends," including those who had helped the Mongols in their military campaigns, such as the Kyrgyz, the Uighur Turks and the Persians: those who belonged to this second social group were known as semujen and could trade, use weapons and occupy government posts. The Polos also belonged to this group, and were free to move about the four corners of the Empire, doing business or aiding the Mongol troops in their sieges of enemy castles and cities.

The third rung of the social ladder was made up of the hanren, that is, the inhabitants of northern China, the original Catai. They were allowed to carry on the cultivation of the land, to barter animals, and to come to the city on their rickety carts to sell fruit and vegetables from their fields.

And, finally, the last, the lowest of the lowest rank on the social scale, was that of the nanrem, the people of South China, those poor subjects of the old Song dynasty, the last people to be conquered and subjugated. The condition of the nanrem was truly miserable: they had no right to carry any weapon, they were forbidden to speak the Mongolian language, they could not contract mixed marriages, and their existence basically translated into that of poor slaves. Each Mongolian head of the family had a dozen or so Chinese families under him, exploited for every kind of daily need, for the supply of food, the transport of foodstuffs, the cleaning of the courtyards or the care of animals.

The Venetians used to follow the emperor and his travelling court in their various wanderings to the four corners of the kingdom. Every year, towards the end of August, the emperor was expected to attend a sacred ceremony organised by the shaman sorcerers in the steppes of Mongolia: it took place in the valley where thousands of snow-white mares were bred and was a traditional ritual that the emperor could not possibly miss. Arriving on that plateau with his personal guard of twelve thousand warriors, surrounded by the princes of blood, vassals and other courtiers, the Great Khan sprinkled a little milk from those mares on the ground as a propitiatory rite for the good of everything that belonged to him: men, horses, birds, countryside, cities and castles scattered throughout the mountains. No one could drink of that milk except the direct descendants of the great Genghis Khan and those who had saved his life.

After that ceremony, if the sky was clear of clouds and the air was still the sweet air of summer, the emperor allowed himself a last, majestic hunting expedition: on the back of four elephants was fastened a large cabin covered with tiger skins where, lying on a bed of soft cushions, Kublai Khan enjoyed watching his griffins circling in the sky before they swooped down on their prey. The air resounded with the growl of the gigantic mastiffs sent to track the footsteps of bears, deer and boars, while eagles circled in search of hares and small deer. Thousands of beaters would advance through the bush waving drapes and striking shrubs with spears and sticks, and for all that great agitation, thousands of birds—partridges, turtledoves, pheasants, cranes, herons—would rise in flight, obscuring the sky with their wings.

Those phantasmagorical hunting scenes were followed, in the evening, by equally spectacular parties at which wine made from rice and enriched with fragrant herbs flowed copiously: served in large gold and silver cups, just a few sips of the delicious drink were enough to trigger euphoria among the thousands of diners.

If the festivities fell in the year of the horse, considered the luckiest year of the Mongolian calendar, the celebrations ended up being more wild than usual, with great drunkenness thanks to the consumption of kumys, that sour but equally irresistible liquor made from fermented milk. While jugglers, musicians and other conjurers performed magic tricks, the tables were served large bowls of lotus grain soup, generous courses of apricot goose, and then, in great abundance, all the game of the day: the Chinese servants were constantly shuttling from the kitchens to the tables, carrying on their shoulders long skewers of partridge, pheasant, duck and turtle doves, with the smell of those meats mingling with that of the familiar roast mutton.

And then, finally, it was time to return to Cambaluc.

The Eagle Question

As conceived by Dante, the *Commedia* is intended to be an anticipated Last Judgement, in the sense that justice is the spirit that animates and binds all of the various episodes of this extraordinary journey through the afterlife. Justice is understood as a pure idea and civil passion, but it is also governed by the deepest religious sense. All of the anger for the exile he suffered, the hatred and the deep contempt for a pope like Boniface VIII, the hatred for his direct enemies do not appear as moments of personal revenge, but as episodes of justice that has been violated, offended and trampled upon.

Dante has destined, in *Paradise*, the highest heaven of Jupiter for the exaltation of justice. Having risen from the reddish light of the sky of Mars, the Poet enters the silvery and white sphere of Jupiter to meet the just spirits. Their "sfavillar" gives rise to geometrical figures, as happens with flocks of birds when they soar merrily in flight. The righteous spirits, singing and always moving in tune with each other, draw large letters of the alphabet in the sky, spelling out "Diligite iustitiam qui iudicatis terram," love justice, you who are judges on earth. Then, gathering all together on the final M of "terram" to sing the praises of God, they brighten up like countless flames, until they have transformed that letter into an eagle, the sacred symbol of justice.

Dante is all alone in front of the eagle with open wings, a man who looks with wonder and amazement at that image projected onto the vault of heaven. And he openly expresses one of his doubts, a question of great import at the time: why does God, whose essence is supreme good and supreme justice, condemn for eternity those just and honest souls who, through no fault of their own, did not know faith in Christ? What sin did those who were born on the bank of the Indus commit? Or those who were not baptized by their own choice?

Where is the justice that condemns him thus?
Where is his guilt, if he does not believe?"

Paradiso, XIX

The eagle's reply opens with the solemn, biblical language of a creator God who draws the boundaries of the world and the very limits of human knowledge. The eagle explains that Dante, as a man, cannot set himself up as a judge of such a profound question, nor claim to see with his limited vision a truth that is a thousand miles away:

Well, who are you to sit there on your throne,
acting the judge a thousand miles away,
eyesight as short as some mere finger span?

Paradiso, XIX

At the end of his speech, the eagle begins to circle around Dante like a stork that has just fed its young, while the Poet looks at it in admiration. The eagle sings a song, proclaiming that Dante will not be able to understand it, just as humanity cannot understand divine justice. It is a very delicate question and one that raises profound doubts about the correctness of divine judgement, so Dante compares his desire to know to a fast that has tormented him for years owing to his inability to find food with which to feed himself. The eagle responds with a complex discourse, similar to a scholastic demonstration: in the first part, he reaffirms the imperfection and limitation of human reason in the presence of divine reason, and in the second part, he declares that God's justice is inscrutable, and therefore the human intellect cannot expect to penetrate its secrets, but must accept the truths of faith as they are declared by Scripture.

It is therefore not a true explanation, but rather a stern warning not to be proud like Lucifer, who wanted to rebel against his Creator to become his equal. The fitting answer of the eagle, in the iron logic of a theological matrix, is the type of peremptory one that can be expected from fourteenth-century theological knowledge: all of the judgments formulated by men in the course of their lives will be defeated by divine wisdom and justice, which are never fallacious.

Some interpreters have seen in Dante's question to the eagle a reference to Buddha, a figure exalted by Marco Polo in *Il Milione* as a supreme example of honesty. Marco was so struck by this figure that he expressed the opinion that, "if he had been a baptized Christian, he would have been a great saint to God." And he went on to add: "And yes, I tell you that idolaters from the farthest parts come here in pilgrimage, just as Christians go to San Jacopo in Galicia."

When Dante uses the phrase "beside the Indus," it is, in fact, a way of designating the most distant countries commonly used in ancient and medieval literature: although the question has been debated repeatedly in medieval theological casuistry, Dante's vague mention of the one who might have been, according to Marco Polo, "a great saint appo God," even though he was, according to him, the head and father of all Oriental idols, cannot be excluded. This would be the only vague indication of some Dantesque inspiration in the *Comedy* that was taken from *Il Milione*.

The Discovery of Asia

During the course of the year, Kublai Khan usually alternated among his three residences: besides the summer residence and the hunting residence (situated in a northern location overlooking the Yellow Sea), his main residence was Cambaluc, today's Beijing. It was a city like no other in the world, ten or twenty times bigger than Venice and all the rest of the cities that Marco had seen over the course of his life: it had about a million inhabitants, distributed throughout twelve districts, each with its own gate.

The alleys of those quarters, called hutongs, were swarming with people: every day, more than a thousand carts full of silk entered the twelve gates of the city, followed by carts full of all the good things from the surrounding countryside, such as rice, rhubarb, fragrant herbs, pomegranates, plums, pistachios and walnuts. Several farmers carried by hand dozens of cages full of wild animals, such as sable, bats or pangolins, which they sold at the corners of the alleys. The whole thing seemed like a great gigantic ant-hill, all of the ants rushing about in pursuit of some occupation: some pushed a cart, some carried large baskets of wood on their shoulders, some shouted at people to make way for their master's carriage.

The main streets of Cambaluc were straight and intersected at right angles, but all around them was a dense network of small streets and passageways lined with siheyuan, graceful wooden and brick houses surrounded by courtyards and gardens. At the center of this immense urban labyrinth stood the so-called "forbidden city," a large complex of imperial palaces protected by three circles of walls and tens of thousands of soldiers.

The residence of the Great Khan stood out for its size and opulence: the shimmering yellow of the roof was accompanied by the richness of the decorations, the dozens of carpets and banners hanging on the walls, the images of dragons hanging from the ceiling; there were endless rooms, full of large stoves, beautiful coloured vases, and teak cupboards. Behind the main entrance, there was a variegated garden full of lotus flowers, peonies, azaleas, lilies and orchids, enlivened by the sound of fountains and the sight of small lakes surrounded by bamboo canes. Along the avenues, the thick foliage of evergreen trees, some of them gigantic, stood out. The emperor had had them uprooted and transported from afar with the help of elephants. Kublai Khan had ordered that gazelles, fallow deer and deer should roam in that park, and had built a hill covered with lapis lazuli dust.

In the four distinct wings of his palace were the rooms where his four wives lived, with all of their retinues. The astrologers of the court were daily questioned as to the disposition of the stars, as well as about wars, pestilences, earthquakes, and every possible calamity that might alter the sweet course of life. A great bell, placed on a tower a hundred feet high, tolled the time at Cambaluc: its last peal in the evening announced the end of the day and the curfew after which it was prohibited to go about the streets at night.

Marco was enraptured to have borne witness to all of those wonders, so different from those that had struck him as a boy in Venice. They were the beauties of which both his father and his uncle had spoken to him at length, but he was as astonished as an astronomer who, having calculated, through complicated methods, the existence of a new planet, then finally sees it materialize before his eyes, as bright as a diamond. He was in the beating heart of the Great Khan's empire, and he could also boast of enjoying the emperor's esteem.

In the years when the Polos were on their way to Kublai Khan's court, along the dusty and impervious roads of Asia, the Mongol emperor had finally also succeeded in conquering South China, one of the most developed areas in the world: this could be seen from the development of poetry and arts, from the magnificent ink and colour watercolours, from the printing of books (well ahead of the German Gutenberg), from the various technological inventions, such as the compass, gunpowder or the water clock. One of his best generals, with the epic name of "Hundred Eyes," had besieged and conquered all of the major cities of South China, one after the other, crushing all resistance from the Song dynasty: their last emperor, a child of just four years, was installed by "Hundred Eyes" in a Buddhist monastery, where he remained for the rest of his life, dying forty-seven years later.

With the conquest of the whole of China, there was a profound metamorphosis of the Mongol Empire: Kublai Khan placed the acquired refinement of southern China alongside the Mongols' ancient spirit of nomadism, opposing the roughness of the old warriors from the steppe with his new overbearing desire for the luxury and pleasures in which the rulers of southern China had lived for centuries.

Thus began the Yuan dynasty: Kublai Khan tried carefully to preserve the rough traditional Mongol liturgies of power, inherited from the great Genghis

Khan, but was ready to amalgamate them with the needs of the new kingdom, which demanded the stability and permanence of the court, efficiency of the state apparatus, the systematic collection of taxes from the subject peoples, the construction of roads and bridges, and ease of trade. Thus, for example, came about the invention of paper money, which replaced the heavy gold and silver coins and the bartering effected with salt, shells or porcelain: the rectangular banknotes, made out of mulberry bark, were printed with the red stamps of the emperor; whoever dared to forge them knew that he would be put to death instantly. Then, there were the highly efficient mail stations. There were the ruthless tax inspectors of Persian origin, careful to bring Kublai Khan all that was due to him from the subjugated provinces. There were the hydraulic engineers, in charge of building canals that would unite, via water, the North and the South of the country, and of devising embankments suitable for preventing the flooding of the Yellow River.

Marco soon became one of the emperor's most trusted advisors: when he returned from his missions, he would delight the Khan with his wonderful stories about the places and peoples he had visited. In his words were evident all of the acumen and attention to detail regarding the customs and habits of the places that he visited typical of a wise man. He spoke of the production of wine, but also of the beauty of the women, of the gigantic watchtowers of the cities and the warmth of the peasants' houses, of the beauty of the gems and the smell of the sea, of the rustling of the forests and the silence of the deserts, of the ruined castles and the caravanserais, of the masks of the sorcerers and the weapons of the knights, of the scent of ginger and, coming full circle, the inebriating taste of wine seasoned with a thousand spices.

Everything was carefully detailed, thanks to that very keen spirit of observation that the Khan had recognized the first time Marco had appeared before him. On each return to Cambaluc, Marco was received in the great throne room: the Khan sent everyone away and closed his eyes to hear tales of the places in his kingdom that he had not—nor would ever—visit.

So, he imagined the very wide river that crossed the city of Chengdu, over which a stone bridge had nevertheless been built, housing dozens of stalls and wooden shops full of fruit, grains, soybeans and textiles; the crackling of bonfires fed with bamboo canes that scared the beasts away from the bivouacs; the great saltwater lake full of cormorants, where milk-colored pearls were fished; the necromancy of the Tibetans, capable of unleashing storms with thunder and violent lightning; the ferocity of the crocodiles and the poisonousness of the Yunnan snakes; the woods full of game and the regal step of the musk deer; the men and women of Jiangsu who covered their teeth with very thin gold plates and covered the skin of their bodies with a thousand tattoos. Surrounded by the heat of the braziers, enveloped in the fragrant smoke of the resin, Kublai listened and dreamed, closed his eyes and laughed: Tengri, the god of the Mongols, had been good to him, sending him such a clever boy.

Medieval Genius

The astrolabe, the very symbol of medieval science, was a truly amazing paradox: based on the nonsensical Ptolemaic system, in which the Earth was the center of all creation, it nevertheless managed to be one of the most precise and versatile astronomical instruments ever built, capable of calculating the position of celestial bodies such as the Sun, the Moon, the planets and the stars above the horizon or at the zenith, and of determining the local time simply through the latitude, or vice versa.

How this device, as large as a pocket watch and consisting of a couple of circles and indexes, could enclose the representation of the celestial sphere in such a tiny space is one of science's most important questions. So much genius of the medieval world, the sum of ancient civilizations that interpenetrated and enriched each other, represented in one device. The acumen at the base of this instrument went hand in hand with its beauty: sometimes, it shone only with the colours of gold or silver, while, very often, it was also decorated with precious gems arranged along its case.

The invention of the astrolabe is attributed to Hipparchus of Nicaea, who lived in the second century B.C.E., one of the leading experts on the complicated theory of epicycles at the base of complex calculations on the motion of the planets. The operating principle was that of stereographic projection of the firmament, a refined intuition of the first Greek astronomers that was later taken up by Claudius Ptolemy. From Greece, the astrolabe spread first to Alexandria in Egypt and other Mediterranean ports, then to Byzantium and the Islamic world. There, thanks to the profound mathematical and astronomical knowledge of Arab scientists, the astrolabe met with undisputed success and was developed in all possible variations, from the wooden object used by the mosque astronomer to indicate the direction of Mecca or select the hours of prayer to instruments embellished with rubies and lapis lazuli, destined for princely collections. Vega, Aldebaran, Betelgeuse and many other star names are, by the way, one of the finest legacies of Islamic astronomy. The astrolabe consisted of a graduated circle, like a protractor, hollowed out in the centre to accommodate the other parts of the instrument. It included a thin disk upon which was engraved the projection of points on the celestial sphere at a given latitude, and then the so-called "rete," a complex structure that, by means of its points, indicated the position of the fixed stars. It was completed by a rotating arm fixed to the center that allowed for the measurement of angles. This set of circles and pointers was thus the quintessence of the geocentric system of the world, the one that was also the basis of Dante's cosmology in the *Comedy*.

If the astrolabe was the main instrument of all of the civilizations that faced the Mediterranean, the compass was one of the most beautiful inventions of the millenary Chinese civilization. It was the oldest instrument used for orientation, especially on ships, where it served to identify the route to be taken: in fact, its needle always pointed inexorably north. Compass needles were made from some strange stones discovered by chance. Those small stones had something magical about them: if they were thrown into the air, they would attract or repel each other, depending on their relative orientation, but, once they fell to the ground, they always ended up lining up in one direction, as if they had a life of their own.

Who knows who that unknown sailor was, one of those many unknown geniuses in the history of humankind, who had the idea of placing a magnetic needle on a piece of cork floating on the water contained in a small container… This idea changed the history of navigation, and thus the course of the history itself, forever. There was then a crucial improvement of that instrument, due to the replacement of the floating needle with a rotating needle resting on a pin, all enclosed in a case (hence the name of the device) upon which the windrose was drawn.

This admirable chinoiserie became European heritage in the twelfth century, thanks to Arab and Amalfi sailors, to whom we also owe the curious and legendary attribution of the invention of the compass to a certain Flavio Gioia of Amalfi. In reality, this personage never existed, but it was the misinterpretation of a Latin text that gave life to the myth: the text in question said that the historian Flavio Biondo attributed the invention of the compass to Gioia of Amalfi. The philologist Giambattista Pio, who came across this text, mistakenly interpreted that to mean that the invention of the compass was due to Flavio Gioia from Amalfi! The Amalfitans were, in any case, responsible for important improvements to the compass, which is why the rose on the compass symbolizing north was replaced, from a certain point onwards, by the lily, the symbol of the town of Amalfi, later changed to the letter "T," standing for "tramontana": this was the name that the Amalfitans had for north, the same name as the wind coming from the mountains, from the village Tramontano, behind the town.

La Candida Rosa

Having reached the end of his journey into the Empyrean, Dante presents, in the last four Cantos, the angelic choirs and the "white rose" of the blessed, a prelude to the vision of God. The style and language must be up to the subject, and the beginning of Canto XXX already demonstrates that they are, with the complex astronomical simile of the stars that slowly fade in the morning light, compared to the gradual disappearance of the circles of light, and then the passionate praise of the beauty of Beatrice, so superhuman that it can only be fully enjoyed by God. Beatrice prepares Dante for his ascent to the tenth heaven, where he will be shown the angelic choirs and the blessed with the mortal body that they will materially regain—according to the miracle of the resurrection of the flesh—only on Judgment Day, thus obtaining a truly unique privilege.

In Canto XXXI, Beatrice's place will be taken by Saint Bernard, who will accompany Dante to the final vision of the divine mind. Even the entrance into the Empyrean is solemn: Dante is wrapped in an intense light that, at first, prevents him from seeing, after which, regaining a higher vision, he able to see by degrees the triumph of the angels and the blessed. He is presented with a river of light flowing between two banks full of colorful flowers:

> *a river in full spate,*
> *fire-dazzle-gilded, flowing through verges*
> *painted afresh in colours of wonderful spring.*
>
> Paradiso, XXXI

Beatrice warns Dante that he sees only veiled anticipations of the real image of souls in the Empyrean, due to his inability to sustain the gaze of the true essences. He will then sustain the sight of the Rose of the Blessed, the Mystic Rose:

> *This light became a circle in its form,*
> *extending its circumference so far*
> *as might a belt too generous round the sun.*
>
> Paradiso, XXXI

It is a sort of luminous amphitheatre on whose tiers the blessed sit in their seats, appearing as a lake of light, circular in shape and immense in size. The comparison of the council of the blessed with a rose, not alien to the mystical literature of the fourteenth century, is presented here in solemn form. Dante is drawn by Beatrice "in the yellow of the everlasting rose," and looks clearly at the blessed:

Their faces all were bright with living flame,
their wings of gold, their other parts so white
that snow has never reached to that extreme.

Paradiso, XXXI

Dante thus succeeds in intuiting the general form of Paradise and, looking more closely, in grasping the highest part of the Rose that shines even more brightly in the middle, where the Virgin Mary sits, together with "thousands, angels—feasting, dancing." The light of the heavenly Rose is reflected on the concave surface of the Primo Mobile, which draws its movement and virtue from it and reverberates throughout the other heavens, as a flowering hill is reflected in the water of a lake below. Dante sees the souls of the blessed reflected in the light of the Rose, arranged in thousands of steps.

With this extraordinary, almost psychedelic cosmic vision, Dante solves his main cosmological and theological problem: how to achieve a synthesis between the Greek cosmos and Christian theology, placing God's superior creative function at the centre.

Although the vision of the Empyrean offered by Dante in Cantos XXX and XXXI is in no way representable, we note that, in Cantos XXVII and XXVIII, the Poet had described a new geometry, the Riemannian geometry of our real physical Universe, recognized by several contemporary scientists, such as the physicists Mark A. Peterson and Roman Patapievici. It is obviously not being suggested that Dante was a mathematical genius, able to see what only a mathematician of the rank of Bernhard Riemann could have developed in the nineteenth century on the basis of all of the mathematical advances that had taken place since Newton. But we do want to highlight Dante's intuition for the solution of a theological paradox that risked undermining his *Comedy*. Let's take a look at what this is all about.

The paradox in which Dante risked being caught up was that, in a cosmos in which the centre is represented by the Earth, at whose own centre Lucifer is stuck, the latter, the supreme spirit of evil, becomes the very centre of the whole Universe, even that of the Empyrean and the Rose of the Blessed! This is what Dante has to contend with in the last Cantos of his poem, that is, making sure that God is quite obviously the centre of all creation, even if, so far, his journey has taken place in increasingly large concentric circles around the Earth. Let us follow him, then, in these last, decisive steps.

Crossing the crystalline sky, we enter an invisible, incorporeal, eternal world, realm of the highest brightness and the fastest speed, but also exactly upside down compared to the visible one. The angelic choirs express, in inverse form, the order of the celestial spheres: the choir closest to the centre of the invisible world manages the sphere farthest from the centre of the visible world. Thus, the dimensions of the visible world are the inverted dimensions of the invisible world. As a result, in the transition from the ninth heaven to the Empyrean, we proceed from one system of visible, corporeal concentric spheres, with the Earth at the centre, to another system of invisible, incorporeal concentric spheres, with God at the centre. The upward direction remains constant, while the speed of the spheres, beginning with the orbit of the Moon, continues to increase up until the orbit of the Seraphim.

Dante's unitary world envisages only one "true" centre, that in which the Creator is found. The sphere of the Empyrean must therefore contain the entire visible Universe in its concentric succession of nine luminous circles, the nine angelic choirs of Dionysius the Areopagite, which rotate, at an ever-increasing speed, around an extremely luminous point, God. The Empyrean has a structure of concentric circles comparable to that of the other nine circles of the visible and corporeal world, but it has different centre than that of the visible world. Dante finds himself there in Canto XXX, and it is like a point of light:

> *for ever round the point that conquered me –*
> *enclosed, it seems, by that which they enclose*
>
> Paradiso, XXX

Dante does indeed use the word "circles" and not spheres, but, in this, he once again follows the teaching of the master Brunetto, who, in the *Tesoretto*, describing the heavens, uses "cercle" and "circle,", even though he means the celestial spheres, which he depicts as eggshells, visible from within. Brunetto taught Dante to think about the geometry of a sphere by looking at it from the inside, and Dante came to imagine the celestial spheres from the inside, in the Empyrean "enclosed by that which they enclose."

In Dante's unitary vision, which maintains a single Universe, although one that is distinct in its corporeal and visible and incorporeal and invisible parts, God thus turns out to be, paradoxically, the centre and, at the same time, also the circumference of the world. A geometry that can make us "see" this paradoxical spatial disposition is, as we anticipated, the spherical geometry theorized in 1854 by the German mathematician and physicist Riemann for a four-dimensional space, difficult to visualize in our minds.

For Dante, Riemannian geometry, which Albert Einstein used extensively to formulate his theory of general relativity and to provide an answer to the nature of the force of gravity and the destiny of the cosmos, was an unconscious attempt to reconcile Aristotelian cosmology with the Christian vision of the world, a way of harmoniously bringing together matter and spirit, temporality and eternity. The originality of Dante's vision consisted in his description of a spatial dimension unknown to him, which allowed for the solution to the cosmological problem arising from the comparison between Christian theory and Greek astronomy.

Dante did not "discover" spherical geometry, but his vision of the cosmos and of God respects the scientific constraints posed by the Aristotelian-Tolemaic Universe and, at the same time, the theological constraints posed by Christianity, with a description that is both realistic and mystical. It represents, for the first time, a complete form of the Christian Universe, which contains within itself the caesura between the visible world and the invisible world, between the four elements and the ether, between a speed and an intensity of light that increases on the way from the smallest visible spheres and grows from the highest sphere of the Empyrean to the blazing and most rapid point where God, the source of light, the fastest and most animate of entities, is gathered.

Dante thus succeeds in giving a plastic image of the divergent correspondence between the geocentric and devil-centric visible world and the theo-centric invisible universe. The visible world presents itself as a vortex with fast edges and slower and slower movements as we approach the centre of the Earth, while the invisible world presents itself as a vortex with slow edges and faster and faster movements as we approach the centre, God.

Everything is held in Dante's unitary cosmos, even if an unresolved question remains: if the Empyrean contains within itself all of the celestial spheres, and therefore lacks a centre, how can God be the geometric centre of the Empyrean?

All powers of high imagining here failed.
But now my will and my desire were turned,
as wheels that move in equilibrium,
by love that moves the sun and other stars.

Paradiso, XXXIII

At the end of his crazy journey, enraptured as if in a dream and blinded by a flash of lightning, Dante finally had a vision of God. It was only a moment, but he understood that that light was the limpid wonder of eternity.

The Long Goodbye

Amid the travelling and trafficking, many years had passed since the three Venetians had arrived at the court of Kublai Khan. The monarch had become old and weak, and the Polos felt that the air was changing: it was time for them to return home, so as not to risk being overwhelmed by all the problems, anxieties and showdowns that usually followed the arrival of a new emperor. But Kublai didn't seem at all willing to let them go; Marco's voice still comforted him when they gathered around the brazier in the evening, amidst the scent of camphor and the intense smell of spiced wine: in those moments, he remembered when he ran like the wind over the colts of the steppe.

Fortunately, it was chance that changed the course of things. One fine day, a group of ambassadors from faraway Persia arrived in Cambaluc: Khan Argon, also from the Mongol dynasty, sent them with the task of finding him a bride of the same lineage as Kublai in order to keep the ties between the two Khanates strong.

The chosen one was the beautiful princess Cocacin, whose name meant "Celestial." The ambassadors, together with the young girl, left for Persia, but, in a few months, they were back in Cambaluc: they had had to turn back because the routes of the Silk Road, which they had travelled without any problems on their way there, were now closed due to the outbreak of a war, and they did not feel like risking the princess's life by going into territories overrun by armies.

It was the three Venetians, as we have already mentioned, who had the brilliant idea that the Persians could return by sea, and that they would be happy to accompany them. Reluctantly, Kublai Khan agreed to their departure: he knew that this was Marco's last visit and that he would never again be cheered by his stories. He gave the Persian ambassadors and the three Venetians the traditional golden tablets as a passport to all the lands of his empire, and had an imposing fleet of fourteen junks fitted out. Each of these boats had three main masts and dozens of sails with the image of fearsome dragons, while on board, there were comfortable cabins and plenty of food, even live animals to be slaughtered when necessary. Amidst mountains of baggage, more than a thousand people took their places on the ships, including crew and passengers, dozens of whom were the princess' bridesmaids.

The fleet left the port of Quanzhou in early 1292 and headed south, looking forward to a pleasant journey of a couple of months in the waters of the eastern oceans. But the reality turned out to be much more dramatic: the voyage lasted more than a year and a half, and of the thousand people on board, only eighteen arrived in the Persian port of Ormuz alive. Eighteen… and among those miraculous few were, needless to say, the ever-eternal Polos; aside from them, there were only one of the three original Persian ambassadors, Princess Cocacin, the commander of that ill-fated fleet and twelve sailors.

What were the causes and who was responsible for that carnage? There were many answers to these questions: violent storms on the open sea, dramatic shipwrecks, but also disease, assaults by ferocious pirates and, above all, an atrocious overestimation of those Chinese junks, which were perhaps useful for fishing off the coast or transporting cargo from one Chinese port to another, but were disconcertingly inadequate for oceanic routes and their monsoons.

In the classic style of understatement typical of Marco Polo, his account of such a dramatic journey totally lacks any sort of tragic tone. On the contrary, the final pages of *Il Milione*, dedicated to the places visited while making their way through the waters of the Pacific Ocean, the South China Sea, the Indian Ocean and the Arabian Sea are among the most beautiful and vivid of the whole book.

The first stop after departure was in Vietnam, whose sovereign was a vassal of Kublai Khan, to whom he supplied twenty elephants a year, together with a generous consignment of aloe wood. From an ethnological point of view, the peculiarity of that kingdom was the privilege ius primae noctis that the monarch had generously carved out for himself, in which no girl was permitted to marry without having first slept with the king. Moreover, he had the option of keeping her as his bride if the maiden was to his particular liking. In the spirit of the purely mercantile taste for counting, Marco puts in writing that that monarch had, at the time of their visit, as many as 326 children, a number that was destined to increase rapidly.

The next stop was along the coast of Sumatra, where they had to stop for several months to wait for winds favourable to navigation: months, it is believed, that were particularly distressing, because the natives of that island, as well as nearby Borneo and New Guinea, had a very bad reputation: that of being devoted to cannibalism. There were also rumours that, on the mountains of those lands, there lived primitive men, very aggressive, with long tails and smooth reddish hair: these were, in fact, orangutans, of which there were numerous specimens on that island. In a related, similarly simian story, in the islands' markets, cunning and cruel traders sold embalmed bodies of certain tiny beings, presenting them as dwarfs from India, when, in reality, they were nothing but skinned monkeys.

In those places, Marco saw a unicorn for the first time, very different from what he had imagined. The skin of the beast was similar to that of an elephant, the snout was squat like that of a wild boar, and it had a particular predilection for muddy marshes: it was perhaps the first time a European had seen a rhinoceros up close. The flora of Sumatra was lush with coconut palms, coniferous forests, and a dense tropical rainforest reflected in the crystal clear waters of the Indian Ocean. Marco collected several stems of verzino, a vegetable to be used for dyeing fabrics, but they did not take root once he reached Venice.

The circus of wonders and horrors was further expanded when they visited the Nicobar and Andaman islands in the Indian Ocean, full of snakes but, above all, of cynocephalic beings, men with the heads and legs of dogs who were also devotees of anthropophagy. By contrast, the king of Jade, the island of gems, wore necklaces around his neck adorned with magnificent, precious, priceless stones—topazes, amethysts, sapphires.

As they rounded the southern tip of India at night, low on the horizon, the North Star had reappeared. They stopped again on the eastern coasts of the Indian continent, near Goa, where Marco's attention was caught by shark charmers and treasure hunters: the latter immersed themselves all day long for the purpose of gathering oysters, from which they extracted precious pearls of all shapes and qualities. But what particularly struck him was the shocking use of ritual suicides by fire, in which the king's worshippers would follow him in death, followed by their wives, who were also expected to off themselves when their husbands died.

He noted with some curiosity the cult of sacred cows and the fact that all inhabitants, including the king and his dignitaries, sat on the ground, ideologically justifying this choice by saying that, "we are made of earth and destined to return to the earth." He recorded Indian customs and habits that had been perpetuated over time and that are still in vogue today, such as the fact that the people effected all of the "beautiful and worldly things," including eating, with their right hands, leaving the left hands to perform all of the baser tasks, including "cleaning the shameful parts and the like." He also met the tribe of the Gavi, to whom local history attributed the murder of Saint Thomas the Apostle, and who, because of this crime, were prohibited from visiting the burial place of the saint.

Those Indian places were full of every good thing: cotton, spices, ginger, turmeric, medicinal plants used by gurus, incense, cinnamon. The smells mixed with the colours, and the bazaars were overflowing with gold filigree, silver jewellery, brilliant stones, pearls and leather goods. Marco made diligent notes in case he was given the opportunity in the future to return to those lands to do business.

After dozens of other vicissitudes, the few survivors of that large expedition finally arrived in the sultry port of Ormuz, where they took the route that they had travelled some twenty years before, through the immense territory of Persia, in order to leave the princess Cocacin with Khan Argon, lord of the Levant Tartars. But, when they arrived at his court, they learned that he had died in the meantime, leaving an heir who was little more than a child: it would be his turn, when he came of age, to marry the princess Cocacin and perpetuate the ties between the Mongol dynasties. While there, they further learned that Kublai Khan had also died: that news definitively signalled the end of an era. The moment of detachment is immortalized by Marco Polo's memory and Rustichello's pen in the most romantic of ways:

> *I will not lie to you that Queen Cocacin was so fond of the three ambassadors that he would do anything for them as for her own father, and wept with sadness at their match, when they had to leave her for return to their country.*

They resumed their journey, riding along the roads that separated them from Trebizond, and from there, they reached Constantinople, where they embarked on one of the Venetian galleys that was returning to the lagoon. Venice welcomed them with its own perfumes, the voices of the quays in the port, the edging of the barges, the light on the dome of St. Mark's, the white of the Doge's Palace. They were unkempt, dirty and ragged in their heavy Tartar garments, so transfigured as to be recognizable by no one. Arriving at their old house, near the church of St. John Chrysostom, they knocked at the door, and, at the voice from within shouting "Who is it?", they answered, as if it were the most natural thing in the world, "The landlords."

In the evening, Ramusio tells us, shaved, cleaned of dust and dressed in sumptuous togas of creamy satin, they gave a great feast to celebrate their return. Marco took the three coarse cloth robes with which they had been clad when they landed and, with the help of a knife, began to undo the hems and double seams, after which, as if by magic, jewels began to fall out of those robes in great quantities: rubies, sapphires, carbuncles, diamonds and emeralds.

Twenty-four years had passed: he had left that house as a boy, but was returning to it as a man. The world had changed, and he with it.

Recommended Reading

The World of Marco Polo

Brusegan M., Scarsella A., Vittoria M., Unusual guide to mysteries, secrets, to the legends and curiosities of Venice, Newton & Compton, Rome 2000.

Calvino I., The Invisible Cities, Mondadori, Milan 2019.

Man J., On the Silk Road. Marco Polo and the meeting between two worlds, Giunti, Florence. Milan 2020.

Polo M., Il Milione, written in Italian by Maria Bellonci, research collaboration by Anna Maria Rimoaldi, ERI, Turin 1982.

Power E., Life in the Middle Ages, Einaudi, Turin 1983.

Whitfield S., Life along the silk road, University of California Press, Oakland 2015.

Wood F., Did Marco Polo go to China?, Secker & Warburg, London 1995.

The world of Dante Alighieri

Alighieri D., The Divine Comedy, commentary by Anna Maria Chiavacci Leonardi, Mondadori, Milan 1994.

Boyde P., Man in the Cosmos. Philosophy of nature and poetry in Dante, Il Mulino, Bologna 1984.

Orr M.A., Dante and the Early Astronomers, Gall and Inglis, London 1913.

Patapievici H.-R., The Eyes of Beatrice. What was Dante's world really like?", B. Mondadori, Milan 2006.

Santagata M., Dante. The novel of his life, Mondadori, Milan 2012.

The Tale of the Comedy, Mondadori, Milan 2017.

GPSR Compliance
The European Union's (EU) General Product Safety Regulation (GPSR) is a set of rules that requires consumer products to be safe and our obligations to ensure this.

If you have any concerns about our products, you can contact us on

ProductSafety@springernature.com

In case Publisher is established outside the EU, the EU authorized representative is:

Springer Nature Customer Service Center GmbH
Europaplatz 3
69115 Heidelberg, Germany

www.ingramcontent.com/pod-product-compliance
Lightning Source LLC
LaVergne TN
LVHW010341260326
834688LV00036B/814